INSTRUCTOR'S GUIDE TO

MATHEMATICS

A Human Endeavor

THIRD EDITION

INSTRUCTOR'S GUIDE TO

MATHEMATICS

A Human Endeavor

HAROLD R. JACOBS

THIRD EDITION

W. H. FREEMAN AND COMPANY
NEW YORK

Printed in the United States of America

Fifth printing, 2000

Contents

Introductory Comments   vii

Recommended Books   x

Lesson Plans

Introduction: Mathematics—A Universal Language   1

Chapter 1   Mathematical Ways of Thinking   6

        2   Number Sequences   24

        3   Functions and Their Graphs   40

        4   Large Numbers and Logarithms   58

        5   Symmetry and Regular Figures   74

        6   Mathematical Curves   96

        7   Methods of Counting   114

        8   The Mathematics of Chance   126

        9   An Introduction to Statistics   141

       10   Topics in Topology   161

Answers to Exercises   175

Sources of Illustrations   331

Introductory Comments

There is a memorable scene in the film *Peggy Sue Got Married* (1986) in which the leading character, having traveled back in time to her senior year in high school, stands up in her algebra class and tells the other students that what they are learning will have no value to them in their future life. This is a rather common view of many adults who have taken only courses in algebra and geometry in which few, if any, attempts have been made to connect the abstractions of the subject to everyday reality.

Many years ago George Boehm wrote:

It is a pity that so few people today are given a chance to appreciate mathematics. Somehow the subject has been lost from the liberal-arts course, where historically it held a central position . . .

When mathematics is taught, it is presented mainly as a collection of slightly related techniques and manipulations. The profound, yet simple, concepts get little attention. If art appreciation were taught in the same way, it would consist mostly of learning how to chip stone and mix paints."*

The primary goal of *Mathematics: A Human Endeavor* is to broaden the view of mathematics held by those, like Peggy Sue, whose narrow experiences with it may cause them to question its value. If it succeeds in accomplishing this, it will produce more than" stone-chippers" and "paint-mixers." Clearly, it will not produce artists either, but at least its readers

---

The New World of Math, by George A. W. Boehm and the editors of *Fortune* (Dial Press, 1959).

may gain an appreciation of the art of mathematics and want to learn more about the subject. I think that the book will have served its purpose if, at the end of a course in which it is used as a text, every student will be able to say, as one did, "After taking this class I feel as though, at the start of the semester, I was sort of sheltered. This class opened up a relatively new world to me."

Although a quick glance at the table of contents might suggest a rather sophisticated treatment, the topics chosen are presented in a way that requires minimal mathematical background and maturity. The exercises emphasize inductive thinking and discovery and, because of their strong intellectual interest and practical nature, are of the type that attract students. It is not assumed that students who use the book know how to solve even simple equations or have much experience with geometric figures. An appendix is included for those who are unfamiliar with, or do not remember, several basic topics that were not felt to be a proper part of the main presentation. Reference to these topics is made in the text at the appropriate places.

Although some lessons can be covered in a single day, others may take two or even three days. You may want to plan the assignments in such a way that much of the work can be done in class. Rather than giving a test at the end of each chapter, I grade the work assigned for each lesson. This works quite well and tends to engender a positive attitude toward the subject and a healthy interest in it for its own sake.

The exercises in most lessons are sequential, and so it is best to assign either complete sets or parts of them rather than even or odd problems. For many classes, the Set III exercises should be considered optional. Most are somewhat more challenging than those in Sets I and II, and some students may not be able to do them. A section titled "Further Exploration" appears at the end of each chapter. Although the problems in these sections are intended for individual study by the better students, you may wish to assign some of them to everyone.

Included in this guide are a large number of specific ideas for presentation in class. I have described many of the lessons in great detail in an attempt to convey a feeling for the plans that have worked well for me in the classroom, even to the extent of including samples of the dialogue with my students. If you are a beginning teacher or have never taught a course of this sort before, I hope that these ideas will be of great help to you. although you will surely not want to follow them rigidly. In practice, I am continually making changes in them or replacing them with others. I am sure that, as you try them out, you too will make improvements.

If in teaching this course you have any questions, criticisms, or suggestions, I would be most happy to hear from you. Correspondence should be addressed to me in care of W. H. Freeman and Company, 41 Madison Avenue, New York, New York 10010.

Harold R. Jacobs

Recommended Books

Cundy, H. Martyn, and A. P. Rollett. *Mathematical Models*, second edition. Oxford University Press, 1961.

Davis, Philip J. and Reuben Hersh. *Descartes' Dream: The World According to Mathematics*. Harcourt Brace Jovanovich, 1986.

Eves, Howard W. *An Introduction to the History of Mathematics*, sixth edition. Saunders, 1990.

Gardner, Martin. *The Scientific American Book of Mathematical Puzzles and Diversions*. Simon and Schuster, 1959.

Gardner, Martin. *The Second Scientific American Book of Mathematical Puzzles and Diversions*. Simon and Schuster, 1961.

Gardner, Martin. *Martin Gardner's New Mathematical Diversions from Scientific American*. Simon and Schuster, 1966.

Gardner, Martin. *The Unexpected Hanging and Other Mathematical Diversions*. Simon and Schuster, 1969.

Gardner, Martin. *Martin Gardner's Sixth Book of Mathematical Diversions from Scientific American*. W. H. Freeman and Company, 1971.

Gardner, Martin. *The Incredible Dr. Matrix*. Charles Scribner's Sons, 1975.

Gardner, Martin. *Mathematical Carnival*. Alfred A. Knopf, 1975.

Gardner, Martin. *Mathematical Magic Show*. Alfred A. Knopf, 1977.

Gardner, Martin. *Mathematical Circus*. Alfred A. Knopf, 1979.

Gardner, Martin. *Wheels, Life, and Other Mathematical Amusements*. W. H. Freeman and Company, 1983.

Gardner, Martin. *Knotted Doughnuts and Other Mathematical Entertainments*. W. H. Freeman and Company, 1985.

Gardner, Martin. *Time Travel and Other Mathematical Bewilderments*. W. H. Freeman and Company, 1988.

Gardner, Martin. *Penrose Tiles to Trapdoor Ciphers*. W. H. Freeman and Company, 1989.

Gardner, Martin. *Fractal Music, Hypercards and More Mathematical Recreations from Scientific American*. W. H. Freeman and Company, 1992.

Hoffman, Paul. *Archimedes' Revenge: The Joys and Perils of Mathematics*. W. W. Norton and Company, 1988.

Kasner, Edward, and James Newman. *Mathematics and the Imagination*. Simon and Schuster, 1940.

Kline, Morris. *Mathematics in Western Culture*. Oxford University Press, 1953.

Peterson, Ivars. *The Mathematical Tourist, Snapshots of Modern Mathematics*. W. H. Freeman and Company, 1988.

Peterson, Ivars. *Islands of Truth, A Mathematical Mystery Cruise*. W. H. Freeman and Company, 1990.

Sawyer, W. W. *Mathematician's Delight*. Penguin Books, 1943.

Steinhaus, H. *Mathematical Snapshots*, third edition. Oxford University Press, 1969.

Lesson Plans

Introduction:
Mathematics—A Universal Language

The introductory lesson has two purposes: to present the idea of the universality of the language of mathematics and to make clear to the students that they are expected to *think*. It is probably better to use the lesson as a basis for class discussion rather than as an assignment.

First day

An article titled "Is Anybody Out There?" in the September 1992 issue of LIFE magazine contained an astounding statement from a report by the National Academy of Sciences:

"Intelligent organisms are as much a part of the universe as stars and galaxies; investigating whether some of the electromagnetic radiation now arriving at Earth was generated by intelligent beings in space may thus be considered a legitimate part of astronomy ... It is hard to imagine a more exciting astro-nomical discovery or one that would have greater impact on human perceptions than the detection of extraterrestrial intelligence."

Partly as a result of this report, the United States government in 1992 launched a $100 million project to search for signals from space that prove what many astronomers now believe: that we are not alone. Pictured on page 1 of the text is the dish of the antenna at Arecibo, Puerto Rico that is being used in this project. This dish is similar to a backyard TV satellite dish, but it measures 1,000 feet across and operates with 20 trillion watts of power!

The idea of other civilizations is not a new one. In the fourth century B.C., the Greek philosopher Metrodoros wrote: "To consider the earth as the only populated world in infinite space is as absurd as to assert that in an entire field sown with wheat only one grain will grow." The great theologian C. S. Lewis has gone so far as to speculate that the dis-

1

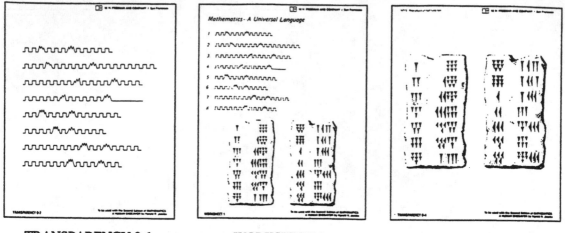

TRANSPARENCY 0-1   WORKSHEET 1   TRANSPARENCY 0-2

tances in space are as large as they are because God, having made many creatures, does not want them to be able to communicate with each other.

After a brief discussion of the appropriateness of mathematics as the basis for beginning a conversation with another world, you might show Transparency 0-1, which contains the radio signal messages from the text.* You may also want to hand out duplicated copies of Worksheet 1 at this time.

Give everyone some time to try to decipher the messages individually before any discussion. The exercises will probably seem very difficult to most of your students. However, after someone has made a "breakthrough" with the first one, they will be revealed as being quite simple after all, and translations of the others will rapidly follow. In some classes, it may be necessary to give the additional hints that the messages are simple mathematical

*Lancelot Hogben suggested the language on which these radio signals are based in a lecture to the British Interplanetary Society in 1952. More information on this language, which Hogben called "Astroglossa," is given in the book *We Are Not Alone* by Walter Sullivan (McGraw-Hill, 1966), chapter 18, "Celestial Syntax."

statements and that they are about numbers. The messages say:

1.	$2 + 3 = 5$	5.	$2 \times 3 = 6$
2.	$3 + 5 = 8$	6.	$4 \times 1 = 4$
3.	$7 - 3 = 4$	7.	$8 \div 2 = 4$
4.	$5 - 5 = 0$	8.	$6 \div 3 = 2$

Transparency 0-2 is for use in presenting the rest of the lesson in accord with the text. In discussing the information on the Babylonian tablet, the class should first explain what the symbols in the left-hand columns of the tablets stand for: 1 through 10, 20, 30, 40, and 50. Some may think that the last four symbols represent 11 through 14; a student should explain why this is unlikely.

After it has been discovered that the first six lines of the first right-hand column read 9, 18, 27, 36, 45, and 54, someone may recognize that this is a multiplication table for the number 9. From the next product in the table, 63, it is evident that whether the symbol $\mathbf{Y}$ stands for 60 or 1 depends on its position in the numeral. The Babylonian numeral system has a primitive place value. Before going any further, you might let everyone try to complete the translation. The numbers in the second right-hand

WORKSHEET 2

TRANSPARENCY 0-3

column are 72, 81, 90, 180, 270, 360, and 450.

Additional information on the Babylonian numeral system can be found in the first chapter of *Episodes from the Early History of Mathematics* by Asger Aaboe, volume 13 of the New Mathematical Library, originally published in 1964 and now available from The Mathematical Association of America.

For something for your students to do as an assignment after the first day's lesson, you might duplicate and hand out copies of Worksheet 2, which contains a modification of an "interplanetary message" devised by Ivan Bell in 1960 and first printed in the *Japan Times*.* As before, most students will probably need help in getting started. Translations of the messages are:

1.   1 1, 2
    1 1 1, 3
    1 1 1 1, 4
    1 1 1 1 1, 5

---

*Martin Gardner included Bell's message in its original form in his Mathematical Games column titled "Thoughts on the Task of Communication with Intelligent Organisms on Other Worlds" in the August 1965 issue of *Scientific American*.

    1 1 1 1 1 1, 6
    1 1 1 1 1 1 1, 7
    1 1 1 1 1 1 1 1, 8
    1 1 1 1 1 1 1 1 1, 9

2. $1 + 1 = 2$
    $1 + 1 + 1 = 3$
    $2 + 1 = 3$
    $3 + 4 = 7$

3. $3 - 1 = 2$
    $4 - 1 = 3$
    $7 - 2 = 5$

4. $1 + 0 = 1$
    $4 + 0 = 4$
    $3 - 3 = 0$

5. $5 + 5 = 10$
    $5 + 6 = 11$
    $10 + 10 = 20$
    $30 + 2 = 32$

6. $2 \times 3 = 6$
    $2 \times 5 = 10$
    $5 \times 12 = 60$
    $10 \times 10 = 100$

1. A A, B
   A A A, C
   A A A A, D
   A A A A A, E
   A A A A A A, F
   A A A A A A A, G
   A A A A A A A A, H
   A A A A A A A A A, I
2. A J A K B
   A J A J A K C
   B J A K C
   C J D K G
3. C L A K B
   D L A K C
   G L B K E
4. A J M K A
   D J M K D
   C L C K M
5. E J E K A M
   E J F K A A
   A M J A M K BM
   CM J B K CB
6. B N C K F
   B N E K A M
   E N A B K FM
   A M N A M K AMM
7. H O B K D
   A B O C K D
   A M M O B M K E

TRANSPARENCY 0-4

TRANSPARENCY 0-5

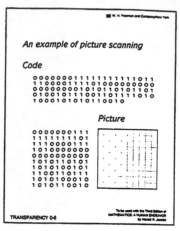

TRANSPARENCY 0-6

7. $8 \div 2 = 4$
   $12 \div 3 = 4$
   $100 \div 20 = 5$

8. $1 \div 10 = .1$
   $1 \div 100 = .01$
   $1 \div 2 = .5$
   $1 \div 4 = .25$

9. $3 \times 3 = 3^2 = 9$
   $4 \times 4 = 4^2 = 16$
   $10^2 = 100$

10. $\sqrt{100} = 10$
    $\sqrt{25} = 5$
    $\sqrt{7^2} = 7$

11. $79.98 \approx 80$
    $99.9 \approx 100$
    $1{,}000{,}000 + 1 \approx 1{,}000{,}000$

12. $\pi \approx 3.1416$

Second day

A good quotation on mathematics as a special kind of language appears on Transparency 0-3.

NASA's "Search for Extra-Terrestrial Intelligence" project is based on the assumption that creatures somewhere in space are sending a message. For us to learn anything from this message, it would have to begin with something very simple, such as counting. The series of statements in Ivan Bell's "interplanetary message" on Worksheet 2 illustrate how one might quickly progress from counting through some simple arithmetic. Transparencies 0-4 and 0-5 are for use in discussing the contents of this message. Following the first part, in which numerals are introduced, the subsequent parts introduce in rapid succession the four basic operations, zero, positional notation and the decimal point, exponents and square roots, approximations and the number pi.

The next step would probably be picture scanning. An example of how the process might work is given in Martin Gardner's article on extraterrestrial communication and appears on Transparency 0-6. The message consists of a numerical code consisting of 100 pulses. The pulses can be represented by the numbers 0 and 1.

Hand out copies of the transparency and

explain that the 100 pulses have been arranged in 10 rows of 10 pulses each to the left of the frame of the picture. Each pulse, or number, corresponds to one of the small squares in the frame. If the number is 1, the square is black and if the number is 0, it is white. The first row of numbers show that only the last three squares in the top row of the picture are shaded.

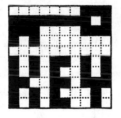

After your students have completed the picture and have discovered that it is a picture of a key and the English word for it, you might point out that picture scanning of this sort is the way in which pictures on television screens and computer monitors are produced. A "picture scanning message" was sent into space from the earth for the first time in 1974. The message consisted of 1,679 pulses and was transmitted from the antenna at Arecibo toward the Great Cluster in Hercules, 25,000 light-years away. Detailed information on the message can be found in the book *Cosmos*, by Carl Sagan (Random House, 1980).

1

Mathematical Ways of Thinking

The first chapter, like the introductory lesson, was written with two goals in mind. Its immediately apparent goal is to introduce the methods of inductive and deductive reasoning by presenting simple examples of each method. Perhaps equally important is the goal of convincing the students that doing mathematics can be an enjoyable and rewarding experience. To accomplish this goal, a variety of puzzles are presented in rapid succession.

It is most important that those students who come to the course with a strong dislike for mathematics due to bad experiences with it in the past be given a chance to overcome this dislike at the very start. An initial successful experience is also important for those who are afraid of the subject because of previous failures in it. The students should be encouraged to discuss their conclusions and results without being told the "correct" answers by the teacher. Try to promote a feeling of mutual discovery and avoid the possible temptation of lecturing the class on all of the assumptions and implications inherent in a particular puzzle. As Voltaire once said, "The art of being a bore is to tell everything."

The first two lessons explore several different questions about the path of a billiard ball on tables of various dimensions. In addition to providing an informal introduction to the methods of reasoning presented in the following lessons, the material has the advantage of putting everyone in the class on an equal basis because it is something that no one is likely to have studied before.

The next two lessons explore inductive reasoning, Lesson 3 emphasizing generalizations that lead to correct conclusions and Lesson 4 introducing examples of inductive patterns that break down and of situations in which appearances are deceiving. The remaining lessons deal with deductive reasoning and its use in formulating simple proofs.

TRANSPARENCY 1-1

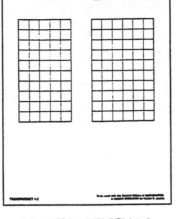

TRANSPARENCY 1-2

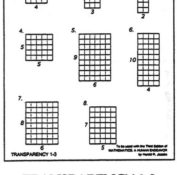

TRANSPARENCY 1-3

The Path of a Billiard Ball

After telling your class that the purpose of the first unit is to introduce them to some of the methods used in developing mathematical ideas, you might show the figure at the top of Transparency 1-1. The modern game of billiards, played with three balls on a table without pockets, originated in the 1870s. In three-cushion billiards, the classic form of the game, a player must hit one ball in such a way that it strikes not only both of the other balls, but also one or more of the cushions at least three different times. (Reveal the diagram of the table below.) For example, a player might plan to hit the black ball so that it rebounds off the lower cushion, strikes the ball at the right, rebounds off the upper cushion and the cushion at the left, and then hits the other ball. If the player is successful, the result might look like this. (Add the overlay.) Needless to say, the game is difficult to master and can be quite exciting to watch.

After this introduction, explain that we are going to analyze the paths taken by just one ball on tables of various dimensions. Transparency 1-2 can then be used to present the lesson in accord with the text.

Be sure to inform your students that they will need an ample supply of graph paper that is ruled either four or five squares per inch or two squares per centimeter for use throughout the course. You might give each student a sheet as a sample, which can be used to do the exercises in this lesson.

If your students begin the assignment in class, you will want to move about the room after a few minutes to detect and help those students who are having difficulty due to confusion or carelessness. The figures for the first eight exercises of Set I are reproduced on Transparency 1-3, which may be useful in helping your class check some of their work.

As mentioned in the Introductory Comments at the beginning of this guide, the Set III exercises should be considered optional for most classes. They are intended for the better students and not everyone will be able to do them successfully.

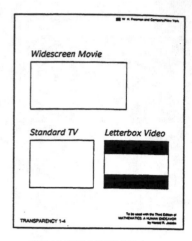

LESSON 2

More Billiard-Ball Mathematics

An interesting way to begin is with a discussion of "letterbox video". This topic reviews the concepts of *ratio* and *similar rectangles*, which are applied in some of the exercises of this lesson.

Have you ever seen a movie on videotape in which the picture does not fill the screen? The black bands above and below the picture give the impression of looking through a mail slot and so the format is called "letterbox."

Movies were originally filmed to produce a picture that is nearly identical in shape to a standard television screen. Many movies are now filmed in a widescreen format instead. Show Transparency 1-4. The figure at the top illustrates the shape of the picture in a widescreen movie. The two figures below show the shape of a television picture. If the movie is shown on television, either only part of the picture can be seen at any given time (shown on the left), or the picture is "letterboxed" (shown on the right.)

The problem is caused by the fact that a widescreen movie and standard TV do not have the same shape. They are not *similar*. The pictures are not similar because their dimensions do not have the same *ratio*.

A television screen that is 18 inches high is 24 inches wide. The ratio of its height to width is

$$\frac{18}{24} = \frac{3}{4} = 0.75$$

A theater screen for a widescreen movie that is 15 feet high is 30 feet wide. The ratio of its height to width is

$$\frac{15}{30} = \frac{1}{2} = 0.5$$

The television and theater screen ratios are not equal and the pictures do not have the same shape.

If this widescreen movie is shown as a letterbox video on a television screen that is 24 inches wide, can you guess how high the picture is? The two pictures have the same shape and so the ratios of their dimensions are

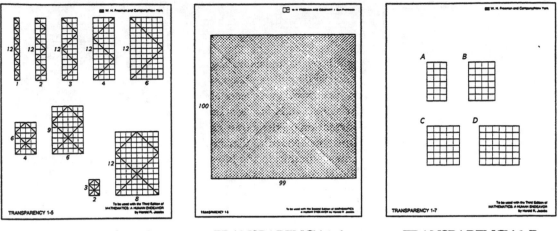

| TRANSPARENCY 1-5 | TRANSPARENCY 1-6 | TRANSPARENCY 1-7 |

equal:

$$\frac{?}{24} = \frac{1}{2}$$

The picture is 12 inches high.

These ideas lead naturally to some brief notes on ratio and similar rectangles as in the text.

REVIEW OF LESSON 1

Transparency 1-3 with its overlay and Transparencies 1-5 and 1-6 are for use in discussing some of the exercises of Lesson 1.

ALTERNATE CLASS OPENER

Hand out duplicated copies of Transparency 1-7 and ask your students to consider the following.

Suppose a ball is hit from the lower-left corner of each table at 45° angles with the sides. On which table do you think the path would be the simplest? On which table do you think the path would be the most complicated?

Give your students time to check their guesses by drawing the paths on all four tables.

Were you correct? Notice that all four tables have the same length: 6. What do the dimensions of the tables have to do with the complexity of the paths? (The greater the common factor of the dimensions of one of these tables, the simpler the path.)

A summary in the following form might be helpful and could be put in the space on Transparency 1-7 below the four tables.

Table	Ratio of dimensions	Greatest common factor
A	$\frac{6}{3} = \frac{3 \times 2}{3 \times 1} = \frac{2}{1}$	3
B	$\frac{6}{4} = \frac{3 \times 2}{2 \times 2} = \frac{3}{2}$	2
C	$\frac{6}{5}$	1
D	$\frac{6}{6} = \frac{6 \times 1}{6 \times 1} = \frac{1}{1}$	6

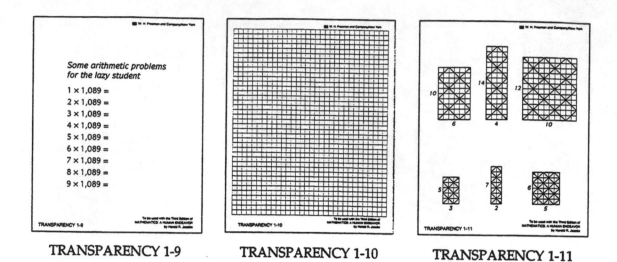

TRANSPARENCY 1-9   TRANSPARENCY 1-10   TRANSPARENCY 1-11

LESSON 3

Inductive Reasoning: Finding and Extending Patterns

After considering the quotation about the person ignorant of mathematical thinking on Transparency 1-8, you might explain that the two lessons about billiard-ball mathematics have been an introduction to the first topic of the course: mathematical ways of thinking.

Hand out duplicated copies of Transparency 1-9. Here is a series of nine problems titled "some arithmetic problems for the lazy student." See what you can discover as you find the answers. Try to use what you discover to be as lazy as possible.

$$1 \times 1,089 = 1,089$$
$$2 \times 1,089 = 2,178$$
$$3 \times 1,089 = 3,267$$
$$4 \times 1,089 = 4,356$$
$$5 \times 1,089 = 5,445$$
$$6 \times 1,089 = 6,534$$
$$7 \times 1,089 = 7,623$$
$$8 \times 1,089 = 8,712$$
$$9 \times 1,089 = 9,801$$

Ordinarily, a calculator would save time in finding answers to problems like these, espe-cially if 1,089 is stored in the calculator's memory. What is it about the answers that makes it possible to guess them? (Each time 1,089 is multiplied by the next larger digit, the first two digits of the answer each increase by 1 and the last two digits each decrease by 1.)

Someone who has observed this pattern and guessed some of the answers by applying it is reasoning *inductively*. There is space above the problems to take some brief notes on in-ductive reasoning in accord with the text.

REVIEW OF LESSON 2

The grid on Transparency 1-10 is for use in discussing the Set I exercises. Transparency 1-11 is for use in discussing the Set II exercises, which consider tables whose lengths and widths are both even.

ALTERNATE CLASS OPENER

Another way to examine the process of induc-tive reasoning is with the Universal Product Code.

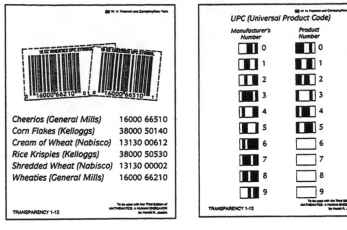

TRANSPARENCY 1-12   TRANSPARENCY 1-13

Show Transparency 1-12. A UPC symbol now appears on almost every item in a supermarket. The UPC, or Universal Product Code, consists of a series of bars of varying widths which represent digits. Some examples are shown here. What do you suppose the digits 16000 represent? (General Mills.) These five digits represent the manufacturer and the next five digits represent the product.

The bars are read by a laser scanner that translates them into the digits of the code. Hand out duplicated copies of Transparency 1-13. The bar symbols used to present the digits of the manufacturers' numbers are different from the symbols used to represent the product numbers. Look at the symbols for the digits in the manufacturer's number. Do all 10 symbols have anything in common? (They all start with a white "bar" on the left and end with a black bar. They all contain exactly two black bars. Of the seven spaces in each symbol, either 2 or 4 are white and 3 or 5 are black.)

Four of the symbols for the product number digits are not shown. Can you guess what they might be? (The product number symbols are the "negatives" of the manufacturer's number symbols.) Rather than letting a student reveal this, you might have four different stu-

dents who have noticed this pattern tell the symbols for the four digits one at a time. The relation between the two types of symbols is connected to the fact that the bars can be scanned in either direction.

Ivars Peterson's remark about mathematics as the science of patterns, quoted as the beginning of Lesson 3, is an appropriate way to tie the Universal Product Code exercise to the lesson on inductive reasoning.

More information on the UPC can be found in the fascinating book, *Reading The Numbers*, by Mary Blocksma (Penguin Books, 1989).

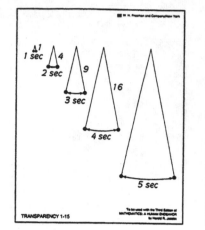

TRANSPARENCY 1-14   TRANSPARENCY 1-15   TRANSPARENCY 1-16

LESSON 4

The Limitations of Inductive Reasoning

Show Transparency 1-14, keeping everything but the top line covered. William Henry Harrison, elected in 1840 as the ninth president of the United States, was the first president to die in office. He caught cold while delivering his inaugural address in freezing rain and died just 31 days later.

Reveal the next three lines. Lincoln, Garfield, and McKinley also died in office—all three were assassinated. Reveal the next two lines. Harding and Franklin Roosevelt also died in office. What pattern do you notice? (Presidents elected at twenty-year intervals [or in years ending with zeros] died in office.) Do you know who was elected president in 1960? (Reveal the next line.)

The pattern is so strong that many people began to believe that it was destined to continue indefinitely. What kind of reasoning is someone using who reaches this conclusion?

Who was elected president in 1980? (Reveal the next line.) Reagan served out both of his terms and broke the pattern. What does that tell you about conclusions obtained through inductive reasoning? (They are never certain.)

REVIEW OF LESSON 3

Transparencies 1-15, 1-16, and 1-17 are for use in discussing the Set II exercises of Lesson 3.

The exercises on Bode's pattern are also a good introduction to Lesson 4. After checking them, point out that there are two more planets beyond Uranus and ask what distances Bode's pattern predicts for this. (384 + 4 = 388 and 768 + 4 = 772.) Neptune, discovered in 1846, has a distance of 300 units and Pluto has a distance of 394 units. It is still not clear whether the numbers in Bode's pattern have any deep significance or whether they are merely a remarkable accident. Martin Gardner discusses it in the "Solar System Oddities" chapter of his *Mathematical Circus* (Knopf, 1979).

Transparency 1-18 illustrates the Set III exercises on the Patterns game by Sidney Sackson.

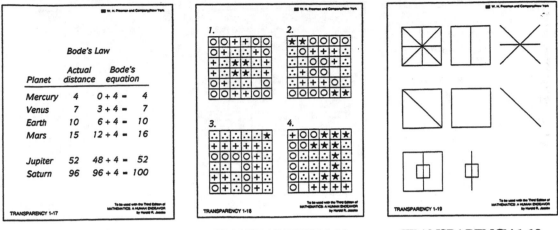

TRANSPARENCY 1-17   **TRANSPARENCY 1-18**   **TRANSPARENCY 1-19**

The game was once featured on the cover of *Scientific American* (November 1969). Detailed information on how it is played can be found in the "Patterns of Induction" chapter of Martin Gardner's *Mathematical Circus* as well as in Sackson's book, *A Gamut of Games* (Random House, 1969).

Mr. Sackson says of the game that it

"... allows, in fact requires, the player to formulate a hypothesis and then test its validity by experimentation. If it stands up, it is probably right. If not, it must be changed to fit the new facts. And the player must be willing, if the experimental data fails to justify it, to throw out a hypothesis completely and come up with a new one. The process is, of course, what we call the scientific method."

Or, as it is referred to on the cover of the magazine, "inductive reasoning."

ALTERNATE CLASS OPENER

Show Transparency 1-19. In each row, the third figure is obtained from the first two by applying a rule. What is the rule, and what figure belongs in the blank space in the third row?*

One way to state the rule is to say that the third figure in each row is found by removing the lines in the second figure from the first figure. According to this rule, the figure that belongs in the blank space in the third row is a large square.

---

*This puzzle is from the September 1978 issue of the French magazine *Science et Vie* and was included in *Fractal Music, Hypercards and More*, by Martin Gardner (W. H. Freeman, 1992).

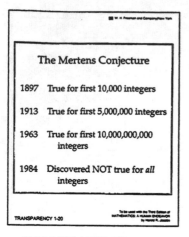

The Mertens Conjecture

1897 True for first 10,000 integers

1913 True for first 5,000,000 integers

1963 True for first 10,000,000,000 integers

1984 Discovered NOT true for *all* integers

TRANSPARENCY 1-20

TRANSPARENCY 1-20

100-Year-Old Hypothesis Disproved

Basic Math Conjecture Found False

A 100-year mathematical conjecture that was known to be true for the first 10 billion numbers has been proved false.

In addition to being a reminder that nothing in mathematics is true until it is proved true, the finding that the so-called Mertens conjecture is false has important consequences in several fields of study, including number theory, combinatorics and algebra.

"It just shows you again that you have to be very careful," Andrew Odlyzko of Bell Laboratories, one of the disprovers of the Mertens conjecture, said Monday. "Empirical evidence can very often be misleading."

TRANSPARENCY 1-21

TRANSPARENCY 1-21

Olympic Games

1896	Athens
1900	Paris
1904	St. Louis
1908	London
1912	Stockholm
1916	
1920	Antwerp
1924	Paris
1928	Amsterdam
1932	Los Angeles
1936	Berlin
1940	
1944	
1948	London

TRANSPARENCY 1-22

TRANSPARENCY 1-22

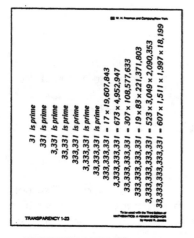

31 is prime
331 is prime
3,331 is prime
33,331 is prime
333,331 is prime
3,333,331 is prime
33,333,331 is prime
333,333,331 = 17 × 19,607,843
3,333,333,331 = 673 × 4,952,947
33,333,333,331 = 307 × 108,577,633
333,333,333,331 = 19 × 83 × 221,371,803
3,333,333,333,331 = 523 × 3,049 × 2,090,353
33,333,333,333,331 = 607 × 1,511 × 1,997 × 18,199

TRANSPARENCY 1-23

TRANSPARENCY 1-23

TRANSPARENCY 1-24

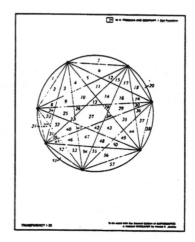

TRANSPARENCY 1-25

TRANSPARENCY 1-25

LESSON 5

Deductive Reasoning: Mathematical Proof

Show Transparency 1-20, keeping everything but the top line covered. In 1897, a mathematician named Mertens made a conjecture about numbers. (Reveal the next line.) Using pencil and paper, he showed that it is true for the first 10,000 integers; that is, that it is true for 1, 2, 3, 4, 5, and so on up to 10,000. (Reveal the next line.) In 1913, another mathematician did the calculations to show that Mertens's conjecture is true for the first 5 million integers. (Reveal the next line.) In 1963, someone used a computer to show that it is true for the first 10 billion integers! As a result of all this work, it seemed reasonable to conclude that the conjecture is true for *all* numbers. What type of reasoning leads to this conclusion? (Induc-

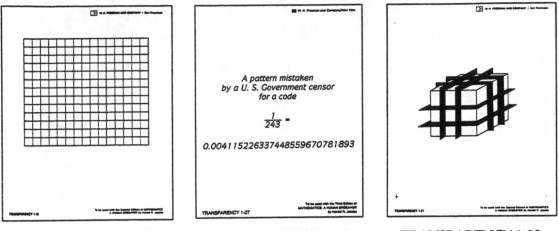

TRANSPARENCY 1-26   TRANSPARENCY 1-27   TRANSPARENCY 1-28

tive.)

Reveal the next line. In 1984, it was discovered that the conjecture is *not* true for all numbers and this surprising result made the newspapers. Show Transparency 1-21 and read the beginning of the article (shown in the enlarged print) to your class. What does this news story remind us about conclusions obtained through inductive reasoning? (They are never certain.)

Unfortunately, the content of the Mertens conjecture would not be meaningful to your students at this point in their study of mathematics. If anyone should ask, explain that it has to do with the numbers of factors of integers, absolute values, and square roots.

REVIEW OF LESSON 4

Transparencies 1-22 and 1-23 are for use in discussing some of the Set I exercises. When showing Transparency 1-22, ask your students why they think the gaps in the table exist. (The Olympics were not held during the First and Second World Wars.) Transparency 1-23 not only reveals the factors of 333,333,331 but also shows the sudden shift from a pattern of primes

to a pattern of composite numbers.

Transparency 1-24 is for use in discussing the first four exercises of Set II. The doubling pattern seems so evident from the first four cases that it is very difficult for most people to believe that they have not made a mistake when they are able to get only 31 regions by connecting six points. An obvious way of checking this is to consider the next case. The figure on Transparency 1-25 reveals that the greatest number of regions that can be formed by joining seven points in every possible way, 57, falls noticeably short of what we would have guessed from the doubling pattern.

Transparency 1-26 can be used with the four pieces from the "cut-outs" section at the end of the transparencies book to illustrate the remaining exercises of Set II.

Transparency 1-27 is for use in discussing Set III. The decimal expansion of

$$\frac{1}{243}$$

reveals the point at which the pattern breaks down. You may want to point out that the pattern holds as long as there is no carrying of digits. It is hidden after that point as shown

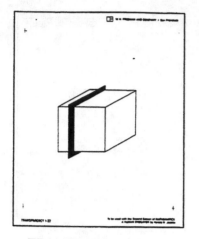

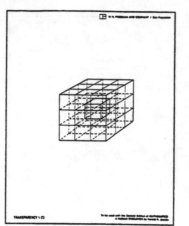

| TRANSPARENCY 1-29 | TRANSPARENCY 1-30 | TRANSPARENCY 1-31 |

here:

$$
\begin{array}{cccc}
 & 10 & 11 & 12 & 13 \\
\ldots 448\ 559\ 66 & 77 & 88 & 99 & \ldots \\
 & \downarrow & \downarrow & \downarrow & \downarrow \\
 & 670 & 781 & 893 & 003
\end{array}
$$

NEW LESSON

An actual Soma cube would be useful in introducing the new lesson. Transparencies 1-28, 1-29, and 1-30 are for possible use in presenting the lesson in accord with the text. Note that Transparency 1-29 should be shown with Overlay A in place first. Then Overlay A is removed and Overlays B and C are added in succession.

The grids on Transparency 1-31, together with a supply of paper clips and pennies, are for possible use in introducing the game considered in the opening exercises of the new lesson.

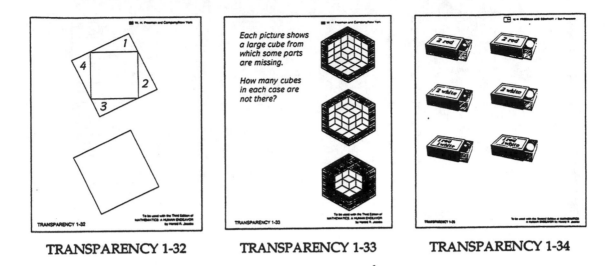

TRANSPARENCY 1-32   TRANSPARENCY 1-33   TRANSPARENCY 1-34

Number Tricks and Deductive Reasoning

A brief presentation of the Pythagorean Theorem is a good way to open the lesson. The proof considered in the Set III exercise of Lesson 5 can be illustrated by using Transparency 1-32 and the accompanying pieces from the "cut outs" section at the end of the transparencies book.

You might also show your students a copy of the book *The Pythagorean Proposition*, by Elisha Scott Loomis, private printed in 1927 and 1940 and available from N.C.T.M. in the Classics in Mathematics Education series. The book contains a history of the Pythagorean Theorem and 370 different proofs of it by deductive reasoning.

ALTERNATE CLASS OPENER

The missing cube puzzles on Transparency 1-33 are from the book *Mind Benders: Games of Shape* by Ivan Moscovich (Vintage Books, 1986). Mr. Moscovich was the creator and original director of the Museum of Science and Technology in Tel Aviv.

Your students may find these puzzles easier to solve if you provide them with duplicated copies. Mr. Moscovich writes of these puzzles:

"These problems depend on our perception of depth, the three-dimensional effect afforded by perspective, in two dimensions. So well is this effect now understood—although either unknown or ignored for millennia before medieval times—that computers can be programmed to recognize three-dimensional objects . . . at any angle, and holograms are not merely works or art to marvel at but used for commercial and security purposes."

He suggests viewing the figures upside down to see the "missing" cubes. The numbers of missing cubes are 26, 19, and 7, respectively. There are a variety of ways to count them which should become evident as your students volunteer their answers and methods.

17

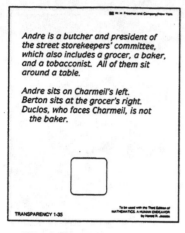

TRANSPARENCY 1-35

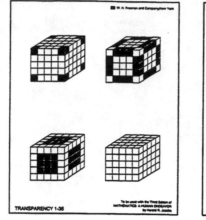

TRANSPARENCY 1-36

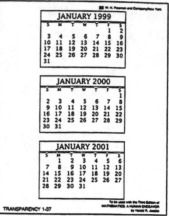

TRANSPARENCY 1-37

Andre is a butcher and president of the street storekeepers' committee, which also includes a grocer, a baker, and a tobacconist. All of them sit around a table.

Andre sits on Charmeil's left.
Berton sits at the grocer's right.
Duclos, who faces Charmeil, is not the baker.

Choose a number.

Add 5.

Double the result.

Subtract 4.

Divide by 2.

Subtract the number first thought of.

The result is 3.

TRANSPARENCY 1-38

REVIEW OF LESSON 5

Transparencies 1-34, 1-35, 1-36, and 1-37 are for use in discussing some of the exercises of Lesson 5.

After reviewing the cube-painting problems of Set II, you might present the general equation. If n represents the number of cubes along one edge, then

$$n^3 = 8 + 12(n-2) + 6(n-2)^2 + (n-2)^3.$$

After observing that n^3 represents the total number of cubes, you might ask what the four terms in the right member of the equation represent.

NEW LESSON

Transparency 1-38 may be used to introduce the new lesson in accord with the text. Have your students try the trick first before showing the transparency and considering the proofs with boxes and circles and with algebraic symbols. The boxes-and-circles proofs are derived from the "bags and marbles" method presented by W. W. Sawyer in his excellent book, Vision in *Elementary Mathematics* (Penguin Books, 1964), now unfortunately out of print. They help the students to visualize the general process and they serve as a guide to writing the proofs with algebraic symbols. If you make up other examples, be careful to design them so that negative numbers and fractions are avoided at every step.

The problem from the Rhind papyrus used to open the lesson is discussed by Richard I. Gillings in the chapter titled "'Think of a Number' Problems" in his book *Mathematics in the Time of the Pharaohs* (MIT Press, 1972).

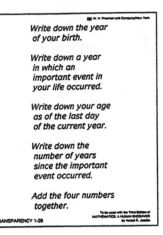

TRANSPARENCY 1-39

Review of Chapter 1

First day

The following number trick, although somewhat complicated to carry out, has quite a surprising result. Tell your students to do the following.

Write down a three-digit number whose first and last digits differ by more than one. (Give examples of both acceptable three-digit numbers and unacceptable ones before proceeding.) Now write down the same three digits in reverse to form another three-digit number. Subtract the smaller three-digit number from the larger one and circle the answer. Reverse the digits in the number that you have just circled to form another number and circle it. Add the two circled numbers.

Multiply the result by one million. (You may want to have someone explain that this is easily done by "adding six zeros to the end of the number.") Then subtract the following number from the result: 827,426,577.* (The number that you give at this point should be written on the board or printed on a card in advance.) Check your work carefully.

Now below each 1 in your answer, write an A, below each 2 write a G, below each 3 write an H, below each 4 write an I, below each 5 write an N, below each 6 write an R, and below each 7 write a T.

---

*The result of the trick is GRANTHIGH, the name of the school in which I teach. The trick is based on the fact that the sum of the two circled numbers is 1,089, so that

$$\begin{array}{r} 1,089,000,000 \\ -\ \ 827,426,577 \\ \hline 261,573,423 \\ \text{GRANTHIGH} \end{array}$$

To adapt the trick to the name of your own school or teams, assign digits to the letters as illustrated in the example below.

BRONCOS	1,089,000
1 0 5 3 4 5 2	− 1,053,452
	35,548

In this example, the students would be told to multiply the sum of the two circled numbers by one thousand, subtract 35,548, and so forth.

Have someone tell his or her result, as some of the students will have made mistakes. You will get an idea of how this remarkable trick works later.*

REVIEW OF LESSON 6

You will probably want to keep the discussion of Lesson 6 very brief. Transparency 1-39 is for possible use in discussing the Set III exercise.

Before the students begin the review exercises, point out that there is a summary at the end of each chapter of the most important ideas introduced in it. These ideas are keyed to the lesson for further reference if necessary.

---

*Your students will learn more about the trick when they do exercises 8-10 of Set II of the review lesson. To prove that the sum of the two circled numbers is 1,089 is a nice, although challenging, problem for second-year algebra students. Let the original three-digit number be $100a + 10b + c$, with $a > c$. Note that $100a + 10b + c = 100(a - 1) + 100 + 10(b - 1) + 10 + c$.

$100(a - 1)$	$+ 100 + 10(b - 1)$	$+ 10 + c$
$- 100c$	$+ 10b$	$+ a$
$100(a - 1 - c)$	$+ 90$	$+ (10 + c - a)$
$+ 100(10 + c - a)$	$+ 90$	$+ (a - 1 - c)$
$100(9)$	$+ 180$	$+ 9 \quad = 1,089$

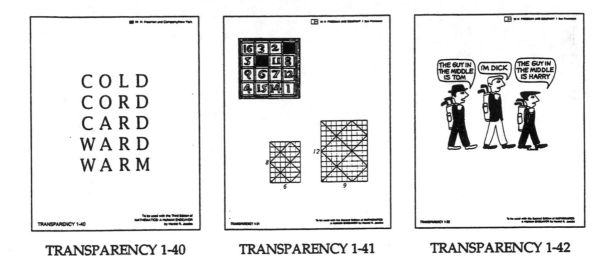

TRANSPARENCY 1-40   TRANSPARENCY 1-41   TRANSPARENCY 1-42

Review of Chapter 1

Second day

Show Transparency 1-40, keeping everything but the top word covered. Then, after telling your students to carefully observe what follows, slowly uncover the next three words one at a time.

What seems to be the rule relating each successive word to the previous one? (One letter is changed.) If you conclude from your observations that there is such a rule, what kind of reasoning are you using? (Inductive.)

Tell your students that there is one more word in the list which makes a logical ending to it and give them some time to try to figure out what it is. When you use the rule, and anything else you have noticed, to figure out the word that logically ends the list, what kind of reasoning are you using? (Deductive.)

Transparencies 1-41, 1-42, and 1-43 are for use in discussing some of the exercises of the review lesson.

The exercise on Dürer's *Melancholy* illustrates the heart of mathematical work. The student must discover a pattern by reasoning

inductively and then determine the missing numbers by reasoning deductively from that pattern. You might mention that the artist included the year in which he made the engraving in the numbers in two adjacent squares and ask if anyone can figure out what the year was. (It is 1514, shown in the second and third squares of the bottom row.)

The five pieces on Transparency 1-43 also appear in the "cut-outs" section at the end of the transparencies book so that they can be placed on the checkerboard grid to confirm the colors of the squares that they cover.

ALTERNATE CLASS OPENER

Here is an alphabetical list of four different alcohols and their molecular weights (Transparency 1-44).

Do you know which one is the alcohol in beer and wine? (Ethanol.) The lightest of the alcohols is poisonous—which one is that? (Methanol.) Propanol commonly occurs in rubbing alcohol.

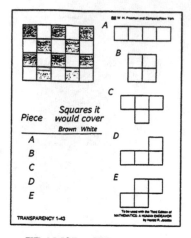

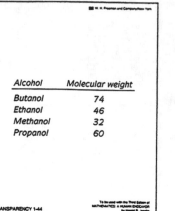

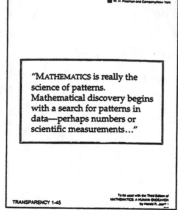

TRANSPARENCY 1-43   TRANSPARENCY 1-44   TRANSPARENCY 1-45

Copy the list but rearrange it in order of increasing molecular weight.

Methanol	32
Ethanol	46
Propanol	60
Butanol	74

The next alcohol heavier than butanol in this series is pentanol. What do you think is its molecular weight? (Each successive molecular weight is 14 more than the previous one, so the molecular weight of pentanol is $74 + 14 = 88$.)

Recall that earlier in this unit, we encountered a description of mathematics as the science of patterns (Transparency 1-45). After noticing patterns, we often draw conclusions from them. In this case we noticed a pattern in the molecular weights of four alcohols and concluded that this pattern might continue to apply to other cases. What kind of reasoning is this? (Inductive.) If we assume that the general conclusion about the pattern in the molecular weights is true, what method of reasoning are we using in finding the molecular weight of pentanol? (Deductive.)

A CHALLENGING PROBLEM

Hand out duplicated copies of Transparency 1-46. The game *Petals Around the Rose* is described in the book *Omni Games*, by Scot Morris (Holt, Rinehart and Winston, 1983).

Mr. Morris writes about it:

"Induction is the logical process of arriving at a general conclusion from a lot of specific facts. Induction is also the process by which one is admitted to membership in a society. Petals Around the Rose is an induction game in both senses of the word.

It is played with five dice. One person rolls the dice, looks at the outcome, and then announces a number. The thrower is Potentate of the Rose and knows the secret. Others, who wish to become members in the Society of Petals Around the Rose, listen to all the Potentate says and try to determine the formula by which he derives the number."

Six examples of throws with the correct outcomes listed in parentheses after them are shown. Can you figure out the secret of Petals Around the Rose?

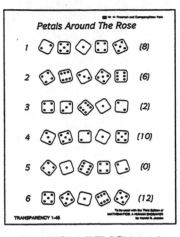

TRANSPARENCY 1-46

This puzzle takes time to figure out. I will not spoil the game for those teachers who are unfamiliar with it by revealing the secret here. (If you are unable to figure out the secret, you may or may not want to share the puzzle with your class!)

CHAPTER 2

Number Sequences

G. H. Hardy referred to the mathematician as a maker of patterns. This chapter explores some of the patterns that can be expressed with number sequences.

The first two lessons deal with arithmetic and geometric sequences: methods for finding the nth term of such sequences and for finding the sum of the first n terms of arithmetic sequences are developed from examples within the exercises. Some of the properties and applications of the binary sequence, including binary numerals, are considered in the third lesson.

The fourth and fifth lessons deal with power sequences, particularly the sequences of squares and cubes. In addition to the geometric application of these sequences, a variety of patterns from number theory are included. The chapter closes with a lesson on some of the properties of the remarkable Fibonacci sequence, together with several examples of its appearances in nature.

LESSON 1
Arithmetic Sequences

First day

In the 400 meter hurdles, the first hurdle is 45 meters from the beginning of the race. The next three hurdles are 80, 115, and 150 meters from the beginning. Where do you suppose the remaining hurdles are located, given that the race includes ten hurdles in all?

After your students have discovered that the hurdles are spaced 35 meters apart and that the remaining hurdles are 185, 220, 255, 290, 325, and 360 meters from the beginning of the race, you might hand out duplicated copies of Transparency 2-1. If the transparency is turned sideways so that the figure is at the top, the solid line represents the track with each unit representing 10 meters. After numbering every 50 meters along the track, vertical segments can be drawn to represent the hurdles as partially shown in the figure below.

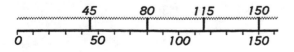

Having done this, your students are prepared to understand the main ideas of this lesson: number sequence, terms of a sequence and the t_n notation, *arithmetic sequence, common difference, and the use of the ellipsis.*

Further examples to illustrate these concepts might include:

a) *A stairs sequence.* Suppose that the first floor of a two-story building is 15 inches above the ground and that the steps up the stairs to the second floor are 6 inches tall:

> 15   21   27   33   39 ...

b) *Driver's license sequence.* Suppose someone gets their first driver's license in 1995 and that the license has to be renewed every 4 years.

> 1995   1999   2003   2007   2011 ...

c) A number sequence that is not arithmetic.

> 10   11   14   19   26 ...

(What seems to be the rule for finding each successive term?)

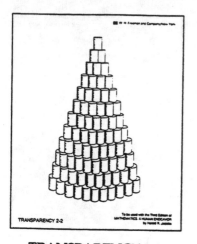

TRANSPARENCY 2-2

d) An arithmetic sequence whose first and fourth terms are 12 and 57:

12 ▓▓▓ ▓▓▓ 57 ...

(How could you figure out the two terms between them?)

Second day

If a second day is spent on Lesson 1, the following problem makes a good introduction to the quick methods for finding the nth term of an arithmetic sequence and the sum of the first n terms.

A fifth-grade class in a Los Angeles elementary school built a "fireproof" Christmas tree from large juice cans which the children had collected and wrapped in green paper. (Transparency 2-2.) When the tree was finished, the teacher posed the following question: The top layer has 1 can, each succeeding layer has 2 more cans, and the base layer has 23 cans. How many cans are in the entire tree?

Let your students solve the problem as the fifth graders probably solved it:

$$1 + 3 + 5 + 7 + 9 + 11 + 13 + 15 +$$

$$17 + 19 + 21 + 23 = 144$$

before looking at the following patterns:

$$
\begin{aligned}
t_1 &= 1 \\
t_2 &= 1 + 2 \\
t_3 &= 1 + 2 + 2 \\
t_4 &= 1 + 2 + 2 + 2 \\
t_5 &= 1 + 2 + 2 + 2 + 2 \ \ ...
\end{aligned}
$$

so

$$
\begin{aligned}
t_1 &= 1 + 0 \cdot 2 \\
t_2 &= 1 + 1 \cdot 2 \\
t_3 &= 1 + 2 \cdot 2 \\
t_4 &= 1 + 3 \cdot 2 \\
t_5 &= 1 + 4 \cdot 2 \ \ ...
\end{aligned}
$$

or, in general,

$$t_n = 1 + (n - 1) \cdot 2$$

The number of cans in the tree written "forwards" and "backwards" is:

$$
\begin{array}{rrrrrrr}
 & 1 + & 3 + & 5 + & 7 + & 9 + \cdots + & 23 \\
\text{and} & \underline{23 +} & \underline{21 +} & \underline{19 +} & \underline{17 +} & \underline{15 + \cdots +} & \underline{\ 1} \\
 & 24 + & 24 + & 24 + & 24 + & 24 + \cdots + & 24
\end{array}
$$

Twelve 24's are 288 but this method counts each layer of cans, and hence each can, twice:

$$\frac{288}{2} = 144.$$

An additional example to illustrate these ideas is suggested by Transparency 2-3. According to the carol, "The Twelve Days of Christmas", the presents my true love gave to me were:

1 partridge
2 turtle doves
3 French hens
4 calling birds
5 golden rings
6 geese
7 swans
8 maids
9 ladies
10 lords
11 pipers
12 drummers

How many presents do these 12 numbers add up to? $(1 + 12 = 13; \dfrac{12 \cdot 13}{2} = 78.)$

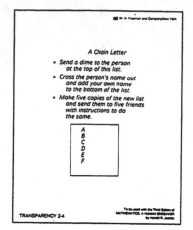

TRANSPARENCY 2-4   TRANSPARENCY 2-5

LESSON 2
Geometric Sequences

First day

You might introduce the lesson on geometric sequences with the example used in the text. If you do, a good way to begin is to give each student a duplicated copy of Transparency 2-4. Explain that the six letters represent the names of six people on a list that you receive, together with the instructions shown. Have your students cross off the letter A from the top of the list and write their name below the letter F. Ask how many lists will have name B at the top (5) and how many will have name C at the top (25). Then have your students figure out the numbers of lists with D, E, F, and finally their name at the top. Note that you do not receive any money until your name is at the top of the list, but that if no one breaks the chain, you will receive 15,625 dimes or $1,562.50! A rather surprising return for an investment of just 10¢ and 5 stamps!

After your students have figured all of this out, show Transparency 2-5 and point out that this chain letter scheme was actually carried out in Denver during the depression of the 1930's.

There is room on the chain letter worksheet (the copy of Transparency 2-4) for your students to make some brief notes on the main ideas of the lesson: *geometric sequence* and *common ratio*.

Further examples to illustrate these concepts might include:

a) *Ancient Egyptian weights*. The units of weight used in ancient Egypt started with a grain, the lightest weight, followed by a shekel, equivalent to 60 grains, and formed a geometric sequence.

Grain	Shekel	Great mina	Talent
1	60	▥▥▥	▥▥▥

It is interesting to note that three units of time, second, minute, and hour, are related by the first three terms of this same sequence.

b) A geometric sequence whose successive terms get smaller rather than larger:

64   48   ▥▥▥   ▥▥▥   ...

The terms of a sequence

An arithmetic sequence:	5	7	9	11	13	...
t_1	5		$= 5 + 2 \cdot 0$	$= 5$		
t_2	$5 + 2$		$= 5 + 2 \cdot 1$	$= 7$		
t_3	$5 + 2 + 2$		$= 5 + 2 \cdot 2$	$= 9$		
t_4			$=$	$= 11$		
t_5			$=$	$= 13$		

A geometric sequence:	5	10	20	40	80	...
t_1	5		$= 5 \cdot 2^0$	$= 5$		
t_2	$5 \cdot 2$		$= 5 \cdot 2^1$	$= 10$		
t_3	$5 \cdot 2 \cdot 2$		$= 5 \cdot 2^2$	$= 20$		
t_4			$=$	$= 40$		
t_5			$=$	$= 80$		

TRANSPARENCY 2-6

Second day

Recent editions of the *Guinness Book of World Records* include a picture of a small horse standing beside a rooster. The horse, not much taller than the rooster, is the result of breeding a series of successively smaller horses from some horses originally discovered in Argentina.

Suppose that the breeders started with horses with a height of 60 inches and that the horses of the next generation were 54 inches tall. If the heights of successive generations of the horses form a geometric sequence, how tall would the horses of the third and fourth generations be?

Some of your students may need help in finding that the common ratio is 0.9 and in verifying that $60 \times 0.9 = 54$ before they are able to figure out the next two terms. (48.6 and 43.74.)

Transparency 2-6 is for possible use in comparing the patterns underlying the nth term of an arithmetic sequence with the nth term of a geometric sequence. I recommend either providing your students with duplicated copies of it or having them copy it.

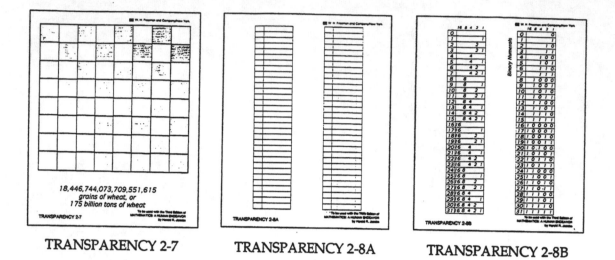

TRANSPARENCY 2-7   TRANSPARENCY 2-8A   TRANSPARENCY 2-8B

LESSON 3

The Binary Sequence

The figure of the chessboard on Transparency 2-7 is for possible use in introducing this lesson as in the text. Keep the lower part of the transparency covered until you come to the end of the story. After illustrating the doubling principle with the numbers of grains of wheat on the first three squares, ask your students to figure out the numbers for the next five squares. Even though the sequence starts out with small numbers, the successive doublings eventually result in such incredibly large numbers that the final sum, shown below the board, is almost beyond belief.

Hand out duplicated copies of Transparency 2-8A and guide your students in making a list of the numbers from 0 to 31 as sums of the first five terms of the binary sequence. Then write the numbers as binary numerals. (See Transparency 2-8B for the expected results.)

ALTERNATE CLASS OPENER

Show Transparency 2-9 and, while you are facing the class, have a student choose a number from 1 to 31 and write it on the board. Then ask the student to sit down and tell you whether or not the number is on each card, reading from left to right. (The *yeses* and *nos* correspond to 1's and 0's in the binary numeral of the number. If a student says, "yes, no, yes, yes, no," the binary numeral is 1 0 1 1 0. Think: 16 + 0 + 4 + 2 + 0 = 22. Do not explain any of this at this point, however.)

Do the trick several times with numbers chosen by different students before explaining that it is based on a number sequence that has many useful properties: the *binary sequence*.

Hand out duplicated copies of Transparency 2-8A and present the rest of the lesson as described above. After your students have completed the two lists, they are ready to understand how the binary number trick works. The numbers are arranged on the grids according to the digits in their binary numerals. Each set of *yeses* and *nos* reveals exactly which terms of the binary sequence can be added to identify the chosen number.

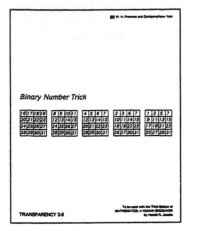

TRANSPARENCY 2-9   TRANSPARENCY 2-10   TRANSPARENCY 2-11

Transparency 2-10 shows part of the first page of an article in *Scientific American* about a series of experiments in which some scientists at the Institute for Behavioral Research in Silver Spring, Maryland, taught two chimpanzees some arithmetic.* The numbers used by the monkeys were in binary form and are shown in the box on the transparency. Three lights were used, the white circles indicating when they were turned on and the black circles when they were turned off.

Transparency 2-11 shows all 64 of the hexagrams from the *I Ching* (pronounced "ee jing") and is for possible use in discussing the Set III exercise of the lesson. More information on the *I Ching* can be found in Chapter 20 of *Knotted Doughnuts and Other Mathematical Entertainments* by Martin Gardner (W. H. Freeman and Company, 1986).

---

*Charles B. Ferster, "Arithmetic Behavior in Chimpanzees," Scientific American, May 1964.

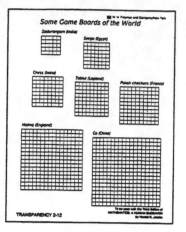

TRANSPARENCY 2-12

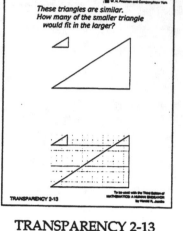

TRANSPARENCY 2-13

LESSON 4
The Sequence of Squares

First day

Throughout history, games in many different countries have been played on boards covered with squares. To introduce the sequence of square numbers, you might hand out duplicated copies of Transparency 2-12 and ask your students to find the number of small squares on each board shown. Having done this, the fact that the square numbers are so-named because they can be depicted by figures such as these boards leads easily into the rest of the lesson: the idea of *exponent*, the meaning of *square root*, and even the relation of squares to sums of consecutive odd numbers. For example, in the case of the Sadurangam board:

$$5 \times 5 = 5^2 = 25,$$
$$\sqrt{25} = 5,$$

and

$$1 + 3 + 5 + 7 + 9 = 25.$$

Information on the games played on the boards illustrated on Transparency 2-12 can be found in The World of Games by Jack Botermans, Tony Burrett, Pieter van Delft, and Carla van Splunteren (Facts on File, 1989).

Second day

Duplicate copies of Transparency 2-13 and fold them in half so that your students first see only the figures on the top half. After recalling that "similar" means "having the same shape," ask your students to write down a guess as to how many of the small triangle would fit in the large triangle. Then have them unfold the paper and use the lower figure to check their guess.

After they have done this, point out that the

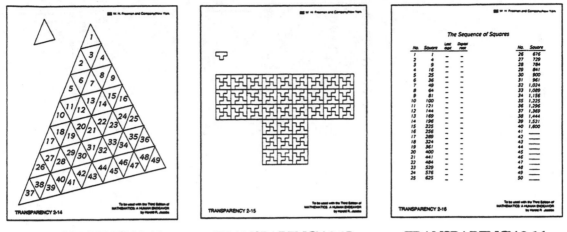

TRANSPARENCY 2-14   TRANSPARENCY 2-15   TRANSPARENCY 2-16

ratio of the corresponding sides of the two triangles is 5:1 but the ratio of the areas of the two triangles is 25:1 or $5^2:1^2$.

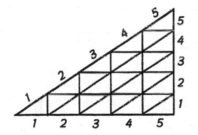

Cover the right column of triangles and observe that, in comparing the size of the remaining figure to the small triangle, the ratio of the corresponding sides is now 4:1 but the ratio of the areas is 16:1 or $4^2:1^2$.. Cover the two right columns of triangles and note the 3:1 and 9:1 ratios.

Show Transparency 2-14, which shows two similar triangles in which the sides of the large triangle are 7 times as long as the sides of the small one. Ask how many copies of the small triangle would fit in the large one before adding the overlay to confirm the answer.

As an example of the fact that this side-area relationship holds for all similar figures and not just triangles, you might show Transparency 2-15, in which the sides of the large T are 12 times those of the small one. The overlay shows one arrangement of 144 small T's filling the larger one.

Duplicated copies of Transparency 2-16 may be useful in considering some of the Set II exercises.

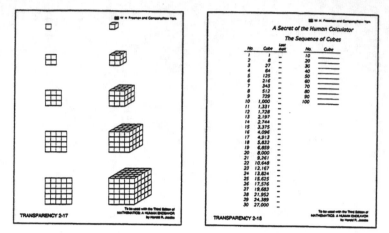

TRANSPARENCY 2-17   TRANSPARENCY 2-18

LESSON 5

The Sequence of Cubes

First day

In 1975, the Hungarian architect and designer Ernö Rubik was granted the patent for what has become one of the most popular toys of the twentieth century. You might begin by holding up a model of Rubik's cube and asking your students how many smaller cubes it appears to contain. Smaller ($2 \times 2 \times 2$) and larger ($4 \times 4 \times 4$ and $5 \times 5 \times 5$) versions of the cube have been created and are illustrated on Transparency 2-17.

Hand out duplicated copies of the transparency and give your students a chance to figure out the number of cubes in the three-dimensional figures before developing the rest of the lesson as in the text.

The definitive book on Rubik's cube is probably *Rubik's Cubic Compendium* by Ernö Rubik, Tamás Varga, Gerzson Kéri, György Marx, and Tamás Vekerdy (Oxford University Press, 1987).

Second day

In his book *Math Magic*, Scott Flansburg reveals many of the secrets of his success as "The Human Calculator" (William Morrow, 1993). One chapter is devoted to finding cube roots. To guide your students in discovering how the secret works, hand out duplicated copies of Transparency 2-18.

First, explain the idea of a cube root with an example: $5^3 = 5 \times 5 \times 5 = 125$, so 125 is the *cube* of 5 and 5 is the *cube root* of 125. Have your students refer to the table of cubes to answer questions such as: What number is the *cube* of 14? What number is the *cube root* of 15,625?

Next, have them fill in the column of the last digits of the cubes and consider questions such as: If a number ends in 3, what digit does its cube end in? If the cube of a number ends in 2, what digit does the number end in?

Condensing this information in a table such as the following may be helpful:

Last digit of number	cube
1	1
2	8
3	7
4	4
5	5
6	6
7	3
8	2
9	9
0	0

Next, have your students fill in the column of the cubes of multiples of 10. Point out that the cube of 100 is 1,000,000, so the cube root of 1,000,000 is 100. It follows that if a number is less than 1,000,000, its cube root is less than 100. Consider some questions such as: If a number is less than 125,000, what number is its cube root less than? If a number is more than 27,000, what number is its cube root more than?

With this background, some of your students may be able to discover the human calculator's secret for doing cube roots. Consider first the cube root of 226,981. If a student is able to recognize that it is 61, let the student explain. If not, ask what the first three digits reveal. (Since 226 > 216, the first digit of the root is 6.) What does the last digit reveal? (If a number ends in 1, its cube root also ends in 1.) So the cube root is 61. Are all six digits of 226,981 used in finding the root? (No.) Which digits *are* used? (The ones before the comma and the last digit.)

For some additional practice, have your students try to figure out the cube roots of these numbers:

1. 970,299 (99)
2. 157,464 (54)
3. 389,017 (73)
4. 681,472 (88)

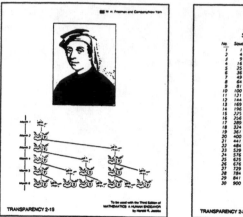

TRANSPARENCY 2-19   TRANSPARENCY 2-20

LESSON 6
The Fibonacci Sequence

First day

If you spend two days on this lesson, you might begin by asking your students if they know what country has had a lot of damage done by wild rabbits. The country is Australia, whose wild rabbit population has been estimated to exceed 200 million, a much larger number than that of Australia's human population. Remarkably, there were no rabbits at all in Australia before the first European settlers introduced them 200 years ago. The rabbits have no natural predators and, because of being able to multiply so rapidly, are costing Australia millions of dollars each year in lost agricultural production.

Following this introduction, your students will be interested in hearing the story of Fibonacci and his rabbits problem as presented in the text. Transparency 2-19 is for use in illustrating this.

After your students have done the first exercise of Set I, you may want to hand out duplicated copies of Transparency 2-20 to make

some of the other exercises easier. For example, your students might circle the Fibonacci numbers that are even before answering exercise 2 and write a check beside each Fibonacci number that is a multiple of 5 before answering exercise 3. The lists of squares and cubes are helpful for exercises 4 and 5 and some of the patterns of Set II.

Second day

An amusing way to start class is with the following number trick, which is based on one of the special properties of the Fibonacci sequence.

Before your students come into the room, put two large cards above the front board, with one beside the other. The more noticeable the cards are, the better. I use cards that are bright yellow.

Remind your students that since the course began, they have done a variety of number tricks. The following trick is especially surprising if your students can keep from making any mistakes in carrying it out.

First, ask your students to write the numbers from 1 through 10 in a column on their papers. Then ask a student to come to the board and write two different numbers on it while you are facing the class. (It is probably safest to specify that each number consist of two digits to prevent your students from being able to carry out the calculations too quickly. Calculators should not be used for the same reason.)

Half the class should copy the numbers from the board on the first two lines of their numbered column. The student who chose the numbers should now erase them from the board, sit down and do the same.

Next, ask a student from the other half of the class to come to the board and write two numbers different from those chosen by the first student. Ask the students in this half of the class to copy these numbers on the first two lines of their numbered column. Again, the student should erase the board and do the same.

Everyone should now carry out the following directions. Add the two numbers and write the sum on the third line. Then add the second and third numbers and write their sum on the fourth line. Continue to do this until numbers have been written on all ten lines, each number being found by adding the two numbers preceding it.

While the class is absorbed in doing this, you must find out what the two seventh numbers are. To do this, ask one of the better students from each half of the class to tell what number he or she has written on the seventh line, saying that number tricks are frequently spoiled by mistakes and that this is a check to see that everyone is on the right track. To find out if these numbers are correct, ask the rest of the class if they agree with them. It is probably best to do this before everyone has had time to find all ten numbers.

Now tell your students to find the sum of all ten numbers. While they are doing this,

multiply the two seventh numbers by 11. Quietly take down the cards from the board, write one result on each card with a felt-tip marker (one that does not squeak!), and put the cards back with the numbers facing the wall. The class should be sufficiently absorbed in adding the ten numbers that you can do this without anyone noticing.

After having your students check their work, ask them if the final result depends on the two numbers originally chosen. Because it does not for many of the tricks we have considered, it is important for everyone to think about this. Then tell the class that you do not know any of the numbers originally chosen and turn over one of the cards to show the number on the back. Then turn over the other card. Those students who have gotten the right answer and who had noticed that the cards were above the board before class will be astonished at seeing their answer on one of them. The fact that the numbers on the cards are different is convincing evidence that the sum does depend on the two numbers originally chosen.

To help your students discover how the trick works, have them represent the two numbers originally chosen as a and b. Have them write these symbols at the right of the original numbers. Explain that all of the rest of the numbers are determined by these two and show how the next three are represented. Then give your students time to figure out the rest themselves.

1. a
2. b
3. $a + b$
4. $a + 2b$
5. $2a + 3b$
6. $3a + 5b$
7. $5a + 8b$
8. $8a + 13b$
9. $13a + 21b$
10. $21a + 34b$

Next, find an expression for the sum of all

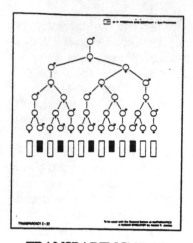

TRANSPARENCY 2-21

ten. (It is $55a + 88b$.) The expression has a simple relation to the expression representing one of the ten numbers in the list. Which one do you suppose it is? (The seventh expression is $5a + 8b$: $11(5a + 8b) = 55a + 88b$.)

If no one has recalled by now that you found out what the seventh numbers were while the class was doing the original arithmetic (I have had many classes in which this had been forgotten), the light should begin to dawn on some of your students at this point. Many, thinking that the cards were above the board all the time, will still be mystified and so this is the time to confess everything.

Transparency 2-21 may be useful in discussing some of the Set I exercises on the family tree of the male bee and the piano keyboard.

More information on the Fibonacci sequence can be found in chapter 13, "Fibonacci and Lucas Numbers," of *Mathematical Circus* by Martin Gardner (Knopf, 1979).

Age	Weight if arithmetic	Weight if geometric
0	8	8
½	16	16
1	24	32
1½	32	64
2	40	128
2½	48	256
3	56	512
3½	64	1,024
4	72	2,048
4½	80	4,096
5	88	8,192
5½	96	16,384
6	104	32,768
6½	112	65,536
7	120	131,072
7½	128	262,144
8	136	524,288
8½	144	1,048,576
9	152	2,097,152
9½	160	4,194,304
10	168	8,388,608
10½	176	16,777,216
11	184	33,554,432
11½	192	67,108,864
12	200	134,217,728
12½	208	268,435,456
13	216	536,870,912
13½	224	1,073,741,824
14	232	2,147,483,648
14½	240	4,294,967,296
15	248	8,589,934,592
15½	256	17,179,869,184
16	264	34,359,738,368

To be used with the Third Edition of
MATHEMATICS, A HUMAN ENDEAVOR
by Harold R. Jacobs

TRANSPARENCY 2-22

TRANSPARENCY 2-22

Review of Chapter 2

Before assigning any of the review exercises, your students will probably need to construct a brief outline of the various types of number sequences. An amusing way to begin this outline is with the fact that the typical new-born baby takes six months to double its weight. (This fact is reported, together with many others that provide good material for creating math problems, in *Never Trust a Calm Dog* by Tom Parker (Harper Perennial, 1990), one of a series of "Rules of Thumb" books that Parker has compiled.)

Suppose that a baby weighs 8 pounds at birth and doubles its weight in six months.

Age	0	½	...
Weight	8	16	...

If the baby kept gaining 8 more pounds every 6 months, how much would it weigh at the age of 4? Continue this sequence to find out.

Age	0	½	1	1½	2
Weight	8	16	24	32	40

Age	2½	3	3½	4
Weight	48	56	64	72

The child would weigh 72 pounds! What kind of number sequence do its weights form? What is the common difference of the sequence?

Now suppose that the same baby that weighed 8 pounds at birth and 16 pounds at 6 months continued to *double* its weight every six months.

Age	0	½	1	1½	...
Weight	8	16	32	64	...

Continue this sequence to find out how much it would weigh at the age of 4:

Age	0	½	1	1½	2
Weight	8	16	32	64	128

Age	2½	3	3½	4
Weight	256	512	1,024	2,048

At the age of 4, the child would weigh more than a ton! What kind of number sequence do its weights form now? What is the common ratio of the sequence?

Your students will probably be astonished at seeing Transparency 2-22, which continues these two sequences to the age of 16.

Functions and Their Graphs

We now turn to the study of simple functions. The students learn in this chapter how, given an equation for a function, to find the coordinates of some of the points in its graph. The graphs of a variety of linear functions, some simple second-degree and third-degree functions, inverse variations, and even a discontinuous function are considered. The use of negative numbers has been kept to a minimum so that students who have had little or no experience with them will not become discouraged.

The emphasis is on discovering patterns and seeing how graphs relate to equations. Students will retain the problem-solving skills and the discovery methods that they learn, even though they will generally not remember these patterns in detail. As in other chapters, a variety of practical examples demonstrate the significance and uses of the concepts being studied.

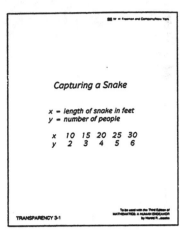

TRANSPARENCY 3-1

LESSON 1

The Idea of a Function

First day

Among the longest snakes in the world are the python and the anaconda, sometimes measuring more than 30 feet in length and weighing several hundred pounds! Capturing such a snake in the wild requires the effort of several people. How many people depends on the length of the snake. (Show Transparency 3-1.) This table suggests a simple rule for estimating the number of people needed. Can you figure out what the rule is?

Length of snake in feet	10	15	20	25	30
Number of people	2	3	4	5	6

(For each five feet of snake, one person is needed.)

Call attention to the fact that this table shows a correspondence between two sets of numbers and use it to develop the central ideas of this lesson: *function, variables, formula,* and *substitution.* For example, label the two sets of numbers x and y and translate the rule for the function into a formula:

$$y = \frac{x}{5}$$

If there were such a thing as a snake 75 feet long, how many people does this formula suggest would be required to capture it? ($y = \frac{75}{5} = 15.$)

Some additional examples showing, given a formula for a function, how to create a table for it may be helpful:

1. Formula: $y = 4x + 1$

x	0	1	2	3	4
y					

2. Formula: $y = 20 - 5x$

x	0	1	2	3	4
y					

3. Formula: $y = (x + 3)^2$

x	1	2	3	4	5
y					

41

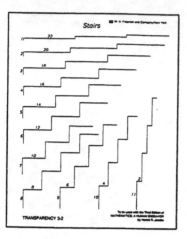

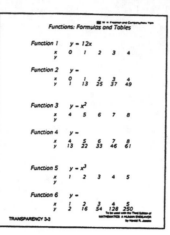

TRANSPARENCY 3-2   TRANSPARENCY 3-3

Second day

There is a staircase in Switzerland that has 11,674 steps and rises 7,759 feet! Approximately how many inches high is each step? $\left(\dfrac{7,759 \times 12}{11,674} \approx \dfrac{8,000 \times 12}{12,000} = 8.\right)$ Each step is about 8 inches high.*

In the 17th century, a French architect developed a formula for designing stairs.

The formula expresses the length of each step from front to back as a function of the height of the step:

$$y = 24 - 2x.$$

Each dimension is given in inches.

Use this formula for copy and complete the following table:

_____

*Many fascinating photographs of stairs are included in the book *Steps and Stairways* by Cleo Baldon and Ib Melchior (Rizzoli, 1989.)

It is interesting to note from the figures on Transparency 3-2 that this stair design formula does not always produce steps of reasonable shape.

Before assigning the Set II exercises of the lesson, you might hand out duplicated copies of Transparency 3-3.

After your students have used the formula for function 1 to complete its table, ask them to guess a formula for function 2 by comparing its table to the table they have just completed. Then give them time to complete the tables for functions 3 and 5 and use them to guess formulas for functions 4 and 6.

An alternate activity might be to play the following game. Tell your class that each student will be asked to name a number to which you will respond with another number. For each number called out by the first few students, respond with the number that is 7 larger. After most of the class has recognized what you are doing, explain that the idea of "adding seven" can be expressed in the formula $y = x + 7$. (A finite set of ordered pairs of numbers does not determine a unique function, but it is highly unlikely that anyone will ask about this. If someone does, you might tell your students that they should look for the "simplest" formula in each case.)

Continue the game with other rules, each time asking your students to guess the rule and express it as a formula. Some appropriate examples are:

$$y = 9x$$
$$y = x - 5$$
$$y = 10x + 8$$
$$y = 20 - x$$
$$y = x^2 + 2$$
$$y = x^3 - 1$$
$$y = x(x + 1) \text{ or } x^2 + x$$

If your students are reasonably successful with this game, you might continue it by having some of them make up the formulas rather than you. Each student who creates a formula should write a table of numbers for it in which the x-numbers are 0, 1, 2, 3, and 4. After the student has read the five corresponding y-numbers to the class, you and the rest of the class can try to figure out the student's formula.

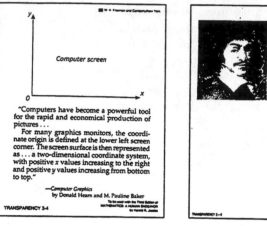

TRANSPARENCY 3-4   TRANSPARENCY 3-5

LESSON 2

Descartes and the Coordinate Graph

The use of computers to produce pictures is now commonplace in a wide variety of applications, including automobile design, special effects in movies, scientific research, video games and medicine.

The figure and quotation on Transparency 3-4 provide a simple introduction to the idea of a coordinate system and, with its use of x values and y values, a connection to the idea of a function introduced in the previous lesson.

Following this, you might show Transparency 3-5 and point out that the idea of a coordinate system was first conceived of several centuries ago. The French mathematician and philosopher René Descartes founded coordinate geometry in 1637 with the publication of his book *A Discourse on the Method of Rightly Conducting the Reason and Seeking Truth in the Sciences.* The portrait of Descartes is an engraving copied from a painting of him by Frans Hals that hangs in the Louvre in Paris. The first page of his work on coordinate geometry is shown at the lower right of the transparency. Interestingly, it appeared as a appendix

to the book which established Descartes as the father of modern philosophy.*

To introduce the rest of the lesson, you might have your students draw a rectangle 15 units wide and 10 units high at the upper left corner of a sheet of graph paper to represent a video screen. Place axes on the rectangle so that the origin falls at its lower left corner and then give your students some point-plotting practice:

Plot the following points and join them in order with straight line segments:

$(4, 5)$, $(8, 6)$, $(7, 10)$, $(3, 9)$, $(4, 5)$.

What figure is formed?

Extend the axes to the left and down to include the other three quadrants. Plot and join the following points in order:

---

*James R. Newman's commentary on Descartes and coordinate geometry in *The World of Mathematics* (Simon and Schuster, 1956, Volume 1, pp. 235-237) is well worth reading.

(1, -6), (1, -8), and (7, -8).

If these points are three corners of a rectangle, where should the fourth corner be? (7, -6).

Plot the following points and label each with its letter name:

A (-4, 7), B (-1, 6), C (0, 5), D (1, 2),
E (0, -1), F (-1, -2), G (-4, -3), H (-7, -2),
I (-8, -1), J (-9, 2), K (-8, 5), L (-7, 6).

On what simple figure do they seem to lie? (A circle.) Draw it. Where do you think its center is? (-5, 1).

You might mention sometime in the lesson that Descartes's name has become part of the mathematical language in that the coordinates of a point in Descartes's system are often referred to as Cartesian coordinates.

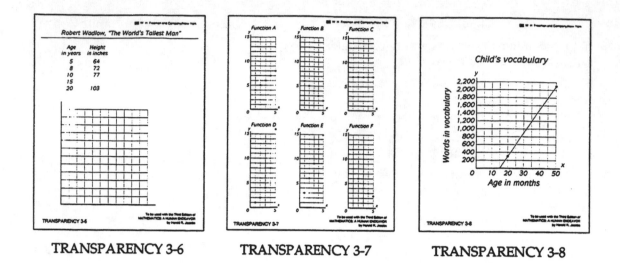

TRANSPARENCY 3-6   TRANSPARENCY 3-7   TRANSPARENCY 3-8

Functions with Line Graphs

First day

The growth of Robert Wadlow makes a nice introduction to this lesson. If you use this example, you might want to hand out duplicated copies of Transparency 3-6 and develop the lesson as in the text. After the students have plotted the four points and discovered that they seem to lie on a straight line, show them how the line can be used to estimate Robert's height at age 15 and at age 22. This abnormal growth started when Robert was 2 and continued until his death at the age of 22.

ALTERNATE CLASS OPENER

You might bring a grocery bag to class with four regular size (26 ounce) containers of salt in it. Before showing your students what is in the bag, ask them if they can guess what ingredient most foods have in common. Salt, as a flavor enhancer or preservative, is in an astonishingly large number of foods, including even such items as corn flakes and butter.

The contents of the four containers of salt represent the approximate amount needed by one person for a year. Many Americans consume several times that amount each year!

Have your students consider the following. Suppose that x = number of days and y = number of teaspoons of salt. A formula expressing the amount of salt you need is

$$y = 1.5x$$

Copy and complete the following table for this function.

x	1	2	4	7	14
y					

Draw and label a pair of coordinate axes extending from 0 to 15 on the x-axis and from 0 to 25 on the y-axis. Plot the five points in the table. What do you notice? (They lie on a straight line.) Extend the line across the graph and label it "amount needed." How many teaspoons of salt does your body need each day? (1.5.) Notice the appearance of this number in the formula.

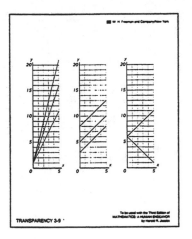

TRANSPARENCY 3-9

A formula for the amount of salt actually consumed by the typical American is

$$y = 5x$$

Copy and complete the following table for this function.

x	1	2	3	4	5
y					

Plot these points on the pair of axes used for the previous function, draw a line through them and label it "average amount consumed." What does the 5 in the formula for this function tell you about the amount of salt consumed by the typical American per day? (It is 5 teaspoons.) What do the two lines for these functions have in common? (They both go through the origin.)

REVIEW OF LESSON 2

Transparencies 3-7 and 3-8 are for use in discussing some of the exercises of Lesson 2.

The formula given in Set III for the size of the vocabulary of a typical child is approximate. It is true during the age interval considered, however, that the rate of growth is almost constant so that the function is close to being linear. The data is from a graph on page 355 of *Introduction to Contemporary Psychology*, by Edmund Fantino and George S. Reynolds (W. H. Freeman and Company, 1975).

Second day

Shoe stores have to carry a much larger stock of shoes than most people realize. It is not uncommon for a store to carry as many as 65 sizes of a single style and to carry as many as 100 different types of shoes.

Your shoe size is a function of the length of your foot. Suppose x = length of foot in inches and y = shoe size. Men's and women's shoe sizes are measured on different scales. For men, the formula is

$$y = 3x - 25;$$

for women, it is

$$y = 3x - 22.$$

Copy and complete two tables for these functions, choosing these values for x:

x	9	10	11	12	13

Plot both functions on the same pair of axes, letting the x-axis extend from 0 to 15 and the y-axis extend from 0 to 20. What do you notice about the result? (The graphs of the two functions are parallel lines.) If a man and a woman have feet of the same length, who has the larger shoe size? (The woman.) (Illustrate this with an example on the graph, such as $x = 12$.) If a man and a woman have the same shoe size, who has the longer feet? (The man.) (Illustrate this with an example on the graph, such as $y = 8$.)

REVIEW OF LESSON 3, Set I

Transparency 3-9 is for use in discussing some of the exercises of Set I of Lesson 3.

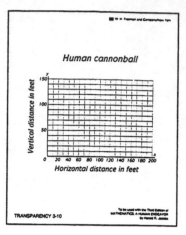

LESSON 4

Functions with Parabolic Graphs

First day

A popular stunt at the circus is the human cannonball. Either hand out duplicated copies of Transparency 3-10 or have your students draw a pair of axes as shown on it. Suppose that the cannon is placed at the origin of a coordinate system and that x and y represent the horizontal and corresponding vertical distances covered by the cannonball in feet. Suppose also that after being shot from the origin, the cannonball goes through the point (20, 35). If there were no such thing as gravity, it would continue along a straight line into outer space. To find some more points on this line, copy and continue this table so that the numbers on both lines form arithmetic sequences.

Plot the three points and extend the line through them.

Because of gravity, the actual table has a more complicated pattern:

x	0	20	40	60	80
y	0	35	60	75	80

It continues:

x	100	120	140	160
y	75	60	35	0

Plot all of these points and then connect them with a smooth curve to show the actual path of the cannonball.

How far away from the cannon does the human cannonball land? (160 feet.) How high in the air does it go? (80 feet.) Emanuel Zacchini, for many years a member of the Ringling Brothers and Barnum and Bailey Circus, holds the world record for traveling the longest distance through the air after being shot from a cannon: 175 feet.

Explain that the path followed by the cannonball is along a curve called a parabola and that many functions have graphs that are this curve.

As a second example of a function with a parabolic graph, you might help your stu-

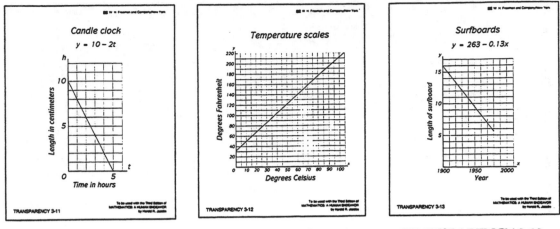

| TRANSPARENCY 3-11 | TRANSPARENCY 3-12 | TRANSPARENCY 3-13 |

dents in constructing a table for the function

$$y = 10x - x^2.$$

Letting x equal the integers 0 through 10 gives

x	0	1	2	3	4	5
y	0	9	16	21	24	25
x	6	7	8	9	10	
y	24	21	16	9	0	

Show how to plot the points on a graph in which each unit on the y-axis represents 2.

REVIEW OF LESSON 3

Transparencies 3-11, 3-12, and 3-13 are for use in discussing some of the exercises of Lesson 3.

Second day

In their book *Fearful Symmetry* (Blackwell, 1992), Ian Stewart and Martin Golubitsky compare the shifting of gears in a car to the changes in gait of a horse:

"When you want an automobile to go faster, you don't always just put your foot on the accelerator pedal and go. Unless the car has an automatic transmission. If not, then every so often you have to change gear. The range of comfortable speeds for the engine isn't as wide as the range of car speeds that the driver may need to use, so a system of gears is employed ... Horses also have a kind of gearing system: they don't always move in the same way. At low speeds they walk; at higher speeds they trot; and at top speed they gallop."

The graph on Transparency 3-14 shows the results of an experiment in which the oxygen consumption of horses is expressed as a function of the speed of the horse.* The graph consists of parts of three overlapping parabolas, corresponding to walking, trotting, and galloping. The numbers on the x-axis represent the speed of the horse in meters per second. The curve on the left shows that if a horse is walking, it consumes the least oxygen at a speed of a little more than 1 meter per second. The second curve represents a horse that is trotting. At approximately what speed does a trotting horse use the least oxygen? (A little more than 3 meters per second.) The curve at the right represents a horse that is galloping. At about what speed does a gallop-

_____

*The oxygen consumption is in milliliters to move 1 meter and the speed is in meters per second.

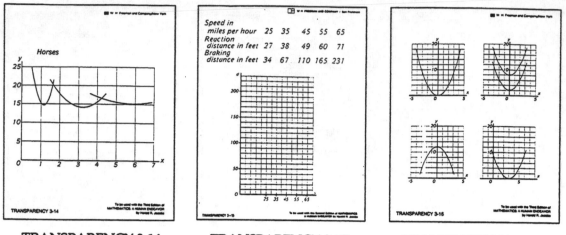

Speed in miles per hour	25	35	45	55	65
Reaction distance in feet	27	38	49	60	71
Braking distance in feet	34	67	110	165	231

TRANSPARENCY 3-14   TRANSPARENCY 3-15   TRANSPARENCY 3-16

ing horse use the least oxygen? (About 6 meters per second.)

ALTERNATE CLASS OPENER

The table on Transparency 3-15 shows the reaction and braking distances of a car as functions of its speed. The reaction distance is the distance traveled by the car between the instant the driver sees a reason for stopping and the instant that the driver applies the brakes. The braking distance is the distance traveled by the car between the instant the brakes are applied and the instant that it comes to a complete stop. The distances given are for a car with perfect four-wheel brakes being driven under ideal conditions.

You might either hand out duplicated copies of the transparency or have your students draw and label a pair of axes as shown. Point out that 1 unit on the speed-axis represents 5 miles per hour and 1 unit on the distance-axis represents 10 feet.

Plot the reaction distances. What do you notice? (They seem to lie along a line.) Draw a line through them. Plot the braking distances on the same pair of axes. What do the points seem to lie on? (A curve.) Draw a

smooth curve through them. The curve is part of a parabola.

REVIEW OF LESSON 4, Set I

Transparency 3-16 is for use in discussing some of the exercises of Lesson 4, Set I.

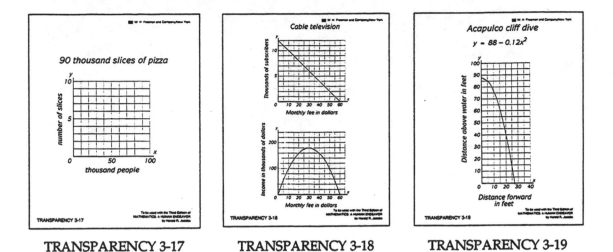

TRANSPARENCY 3-17   TRANSPARENCY 3-18   TRANSPARENCY 3-19

LESSON 5
More Functions with Curved Graphs

First day

In 1987, a pizza with a diameter of approximately 100 feet was baked in Havana, Florida. It was cut into 94,248 slices which were eaten by 30,000 people who had watched it being made. Approximately how many slices did this amount to per person? (3.) Suppose, for simplicity, that the pizza had been cut into exactly 90 thousand slices. Show Transparency 3-17. Copy and complete the following table to show how the number of slices per person would vary with the number of people who shared it.

90 thousand slices of pizza

Thousand people, x	10	20	30	60	90
Number of slices, y	▓▓	▓▓	▓▓	▓▓	▓▓

Can you write a formula for this function? ($y = \dfrac{90}{x}$.) Draw and label a pair of axes as shown on the transparency and graph the function.

REVIEW OF LESSON 4, Sets II and III

Transparencies 3-18, 3-19, and 3-20 are for use in discussing some of the exercises of Lesson 4.

Second day

In his book *Fire*, John W. Lyons writes:

"The fire that burns down a house first burns up one room. In that room it starts small. It goes through a phase of modest growth. Then, in a few moments of very rapid acceleration, it engulfs the entire room and every combustible thing in it.

People are adjusted to linear phenomena. Caught in the first phase of a fire, they expect it to grow steadily and predictably. They are often overtaken by the rapid buildup to flashover: the sudden transition to full involvement of all combustibles in the room. . .

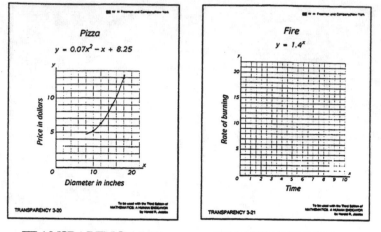

TRANSPARENCY 3-20   TRANSPARENCY 3-21

. . . the production of heat and of other combustible products increases exponentially with time."*

A typical equation for the early growth of a fire is

$$y = 1.4^x$$

in which x is time and y is the rate of burning. The fact that x appears as an *exponent* in this equation is the reason that the fire is said to grow exponentially with time.

What does a graph of this function look like? Either demonstrate with an overhead calculator or have your students use calculators to copy and complete the following table for

$$y = 1.4^x$$

by repeated multiplication.

Time in minutes, x	1	2	3	4
Rate of burning, y	(1.4)	(2.0)	(2.7)	(3.8)

x	5	6	7	8	9
y	(5.4)	(7.5)	(10.5)	(14.8)	(20.7)

*Fire, by John W. Lyons (Scientific American Library, 1985).

Draw and label a pair of axes as on Transparency 3-21, plot the nine points and graph the function by drawing a smooth curve through them.

Having done this, you might go back to quoting from the book *Fire*:

"The production of heat and other combustion products increases exponentially with time. The development of the fire during this phase is proceeding up the steep slope of the curve ... toward the moment when it becomes a room full of flames. Decisions made during this time are life-and-death decisions."

World Record for the Mile

1870	W. C. Gibbs	4:28.8
1880	Walter George	4:23.2
1890	Walter George	4:18.4
1900	Thomas Conneff	4:15.6
1910	Thomas Conneff	4:15.6
1920	Norman Taber	4:12.6
1930	Paavo Nurmi	4:10.4
1940	Sydney Wooderson	4:06.4
1950	Gunder Haegg	4:01.4
1954	Roger Bannister	3:59.4
1960	Herb Elliott	3:54.5
1970	James Ryun	3:51.1
1980	Steve Ovett	3:48.8
1990	Steve Cram	3:46.3

TRANSPARENCY 3-22

TRANSPARENCY 3-22

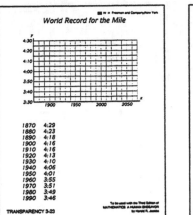

TRANSPARENCY 3-23

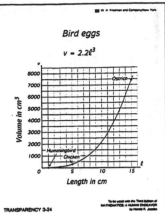

TRANSPARENCY 3-24

LESSON 6

Interpolation and Extrapolation: Guessing Between and Beyond

Show Transparency 3-22 with the lines of the table below 1870 covered. In 1870, the world record for running the mile was slightly less than four and a half minutes. Reveal the remaining lines of the transparency, one at a time. When Roger Bannister ran the mile in less than four minutes in 1954, it created a sensation because many people thought it couldn't be done. Since then, Bannister has predicted that the mile will eventually be run in three and a half minutes!

Hand out duplicated copies of Transparency 3-23 and have your students plot points corresponding to the records shown in the table. Notice that if a line is drawn through the first and last points, all of the other points lie fairly close to it. Extend the line to the x-axis to try to guess approximately when the three-and-a-half minute mile might be run. (Sometime around 2040.)

Explain that in arriving at this result, we are *extrapolating* from the known information. Show how the line can be used to *interpolate* by estimating the records for some of the intermediate years.

REVIEW OF LESSON 5

Transparency 3-24 is for use in discussing one of the exercises in Lesson 5.

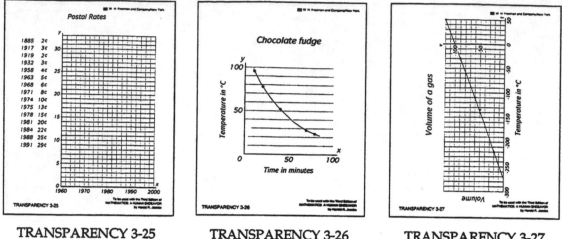

TRANSPARENCY 3-25   TRANSPARENCY 3-26   TRANSPARENCY 3-27

Review of Chapter 3

First day

Show the table on Transparency 3-25 and ask your students what the numbers in it represent. The table lists years in which the cost of mailing a letter was changed and the new costs. (It is interesting to note that the rate was lowered once, something that is unlikely to ever happen again! In 1919, it dropped to two cents after having been raised to three cents two years earlier.)

The postal rate is a function of time. What would a graph of this function look like? Either hand out duplicated copies of the transparency or have your students draw and label a pair of axes as shown. Draw what you think the graph should be from 1950 to the present.*

After plotting the points, most students will probably try to draw a curve through them. Although this is appropriate for showing the general trend in the postal rate, it is not

*There were actually two increases in 1981, but they are treated as a single increase in this table.

really a correct graph of the function: the actual graph consists of a series of horizontal line segments. The graph, unlike others that the students have drawn, is discontinuous. It cannot be drawn without lifting the pencil from the paper. The rate jumps from one value to another without taking on all of the values in between.

REVIEW OF LESSON 6

Transparencies 3-26, 3-27, 3-28, 3-29, 3-30, and 3-31 are for use in discussing some of the exercises of Lesson 6.

In discussing exercises 8-13 of Set I, ask your students if they know the significance of the answer to exercise 12. This temperature is absolute zero and plays a special role in physics.

Second day

Show Transparency 3-32, keeping the line below the figure covered. Some of you may

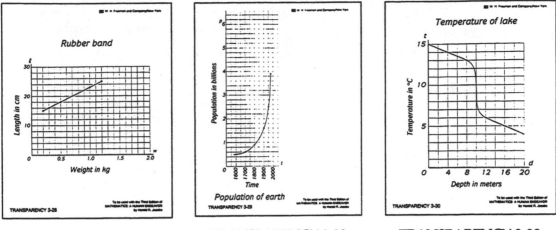

TRANSPARENCY 3-28   TRANSPARENCY 3-29   TRANSPARENCY 3-30

possibly know the following legend about the end of the world:

In the great temple at Benares, beneath the dome which marks the center of the world, rests a brass plate in which are fixed three diamond needles, each a cubit high and as thick as the body of a bee. On one of these needles, at the creation, God placed sixty-four discs of pure gold, the largest disc resting on the brass plate, and the others getting smaller and smaller up to the top one. This is the Tower of Bramah. Day and night unceasingly the priests transfer the discs from one diamond needle to another according to the fixed and immutable laws of Bramah, which require that the priest on duty must not move more than one disc at a time and that he must place this disc on a needle so that there is no smaller disc below it. When the sixty-four discs shall have been thus transferred from the needle on which at creation God placed them to one of the other needles, tower, temple, and Brahmins alike will crumble into dust, and with a thunderclap the world will vanish.*

———————
*Mathematical Recreations and Essays, W. W. Rouse Ball (Macmillan, 1962).

Although this is an amusing story, the puzzle was actually invented in France in 1883 and the story made up to go with it at that time. If you have the puzzle, you might show it to your students. The handsome version, with seven disks, pictured in this photograph is available from Dale Seymour Publications (P.O. Box 10888, Palo Alto, California 94303.)

A popular version of the puzzle contains seven disks. How many moves do you suppose it would take to move the tower from one peg to another? To find out, it is easier to consider a simpler version of the puzzle.

The first time that you do this with a class, give each student a 4-by-6-inch file card, a letter-sized envelope, a compass, and a pair of scissors. Have everyone draw four circles with radii of 3 cm, 2.5 cm, 2 cm, and 1.5 cm on the file card. The circles should be labeled D, C, B, and A, respectively, and cut out. The figure on the overlay for Transparency 3-33 may make these directions clearer. When the puzzle has been solved, the disks can be put in the envelopes and collected for use by other classes. Disks cut from heavy plastic are useful in demonstrating the steps in the solution of the puzzle on the screen.

Remind your students that the puzzle is to

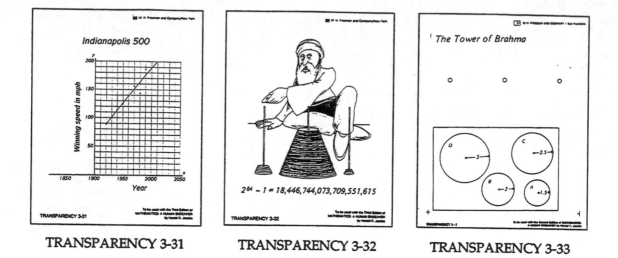

move all of the disks from one peg to another by moving them one at a time and never placing a larger disk on top of a smaller one. Before beginning, each student should draw three small circles as shown on Transparency 3-33 to indicate the positions of the three pegs.

Although our version of the puzzle has four disks, we will begin by using only three. See if you can figure out how to do it. After everyone has had time to try, ask how many moves are necessary. (Seven.) Now try the puzzle with four disks. When you have solved it, do it again, counting the moves. How many moves are necessary? (Fifteen.) If the tower consisted of just one disk, it could be moved in just one step. How many moves are necessary for a tower consisting of two disks? (Three.)

Notice that the number of moves needed to move the tower is a function of the number of disks in it. Record the observations in a table:

Number of disks in tower, x	1	2	3	4
Number of moves needed, y	1	3	7	15

What do you notice? (All the numbers of moves needed are odd. Each number of moves is one more than twice the preceding number.)

How many moves do you think would be needed for a tower consisting of five disks? (31.) Can you explain why this seems reasonable? (It would take 15 moves to move the top four disks to another peg, one move to move the bottom disk to the third peg, and 15 more moves to move the top four disks back on top of it.) Add this to the table and add two more lines to show the numbers of moves that you think would be needed for towers of six and seven disks.

Number of disks in tower, x	5	6	7
Number of moves needed, y	31	63	127

About how much time would it take to solve the puzzle containing seven disks if the disks were moved at the rate of one per second? (About two minutes.)

The tower in the legend consisted of 64 disks. How many moves would that take? It would be nice if we could find the number of moves needed directly from the number of disks in the tower. The sequence of numbers in the second row of the tables is related in a simple way to a number sequence that we studied earlier. Can you figure out which

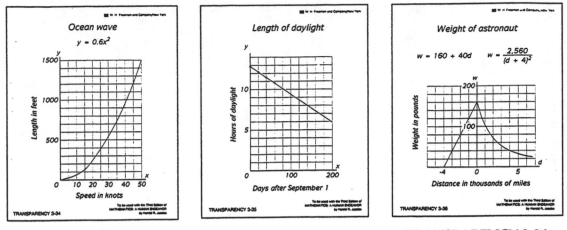

| TRANSPARENCY 3-34 | TRANSPARENCY 3-35 | TRANSPARENCY 3-36 |

one? (Each term is one less than a term of the binary sequence.) List the terms of the binary sequence beginning with 2 in a third row and the terms as *powers* of 2 in a fourth row.

Number of disks in tower, x	1	2	3	4
Number of moves needed, y	1	3	7	15
	2	4	8	16
	2^1	2^2	2^3	2^4

Some of your students may be able to discover from this table that a formula for this function is

$$y = 2^x - 1$$

in which x represents the number of disks in the tower and y represents the number of moves needed. What do you suppose is the number of moves needed to move a tower of 64 disks? ($2^{64} - 1$.) Show Transparency 3-32 again, this time revealing the line at the bottom. This number is incredibly big. At the rate of one move each second, it would take more than *500 billion years.* The universe is thought to be less than 20 billion years old; so even if the legend about the priests in the temple were true, we would have nothing to worry about!

Transparencies 3-34, 3-35, and 3-36 are for use in checking some of the exercises of the review lesson.

Large Numbers and Logarithms

This chapter begins with two lessons on large numbers. In addition to showing how to express large numbers as powers of ten and in scientific notation, these lessons also reveal our difficulty in comprehending the magnitudes of such numbers.

In the ninth edition of the *Encyclopedia Britannica* (1875), J. W. L. Glaisher wrote:

The invention of logarithms and the calculation of the earlier tables forms a very striking episode in the history of exact science, and, with the exception of the *Principia* of Newton, there is no mathematical work published in the country which has produced such important consequences, or to which so much interest attaches as to Napier's *Descriptio*.

Although logarithms are no longer important as a means of computation, their properties will always be useful.* For this reason, the remaining lessons in this chapter consider some of these properties and their applications.

Logarithms are introduced by pairing an arithmetic sequence with a geometric sequence to form a logarithmic function. Binary logarithms are used to illustrate and explain the multiplication, division, and "raising to a power" properties of logarithms. Decimal logarithms are then introduced, the work with them being limited at first to the logarithms of integers. After the relation between the form of a number in scientific notation and its decimal logarithm has been considered, an expanded, but still very simple, logarithm table is then applied to constructing some logarithmic scales. The chapter concludes with a lesson on exponential functions and their applications.

---

*It is worth noting, for example, that, although electronic calculators have made logarithm tables obsolete, every scientific calculator has a logarithm key.

```
3672403
3672404
3672405
3672406
3672407
3672408
3672409
3672410
3672411
3672412
3672413
3672414
3672415
3672416
3672417
3672418
3672419
3672420
3672421
3672422
3672423
3672424
3672425
3672426
```
W. H. Freeman and Company/New York

Counting to a billion

To be used with the Third Edition of
MATHEMATICS: A HUMAN ENDEAVOR
by Harold R. Jacobs

TRANSPARENCY 4-1

TRANSPARENCY 4-1

LESSON 1
Large Numbers

People have all kinds of hobbies. A man who lives in Riverton, Wyoming has one so unusual that it has been written up in newspapers and featured on a television program named "The Best of the Worst."

What is his hobby? Counting! He started with 1 in 1982 and has been spending several hours each day counting ever since. He uses a calculator with a roll of paper in it and types the numbers in order. During the first ten years of doing this, he got to the numbers shown on Transparency 4-1.

When a reporter asked him why he was doing this, he said just because he liked to count and was quoted as saying: "There are probably other things I could do instead of counting but it's an interesting hobby. I don't think I'll get tired of it."*

His goal is to count to one billion and he thinks he will get there by the time he is 62. How big is a billion? It is easy to write 1 followed by 9 zeros, but how big is that? It

---

Los Angeles Times, April 15, 1992.

took him ten years to get to 3,672,426. If we suppose he counted at a steady rate, about how many numbers did he count each year? ($\frac{3,672,426}{10}$ = 367,242.6.) If he continues to count at this rate, how many years will it take him to get to one billion? Show the beginning of the long division on the board:

$$\frac{2,—}{367{,}243)\overline{1{,}000{,}000{,}000}}$$

It would take more than 2,000 years! At the rate he is going, there is no way that he can get to 1 billion by the time he is 62!

Even though we hear the number one billion quite frequently, it is such a large number that it is almost impossible to comprehend its size. Even the number one thousand is bigger than you might think. A week is not a very long period of time. How long is a thousand weeks? Were you alive a thousand weeks ago? It is rather astonishing to realize that a thousand weeks is nearly twenty years!

59

From "One Million"

2
8,399
27,722
46,399
76,115
100,874
116,000
252,055
343,000
451,000
507,635
605,584
774,900
861,500
974,731
1,000,000

TRANSPARENCY 4-2

Following this introduction, you may either want to have your students read Lesson 1 or you may prefer to develop the rest of the lesson as in the text. The concept of building large numbers as products of 10's and the connection between exponential form and decimal form need to be emphasized.

ALTERNATE CLASS OPENER

The book *One Million* by Hendrik Hertzberg is available in paperback (Times Books, 1993). It contains 1,000,000 dots spread over 200 pages with 5,000 dots on each page. Comments explaining the significance of some of the numbers appear on each page. The author writes in the preface to his book:

To "know" something intellectually and to experience it concretely are two very different things. All of us have had concrete experience with the smaller numbers, the ones and tens and hundreds, even the thousands. Millions are harder to come by. Hence what follows.

A few of the numbers in *One Million* are listed on Transparency 4-2; the following comments accompany them.

2	Population of the Garden of Eden.
8,399	Number of times Babe Ruth officially went to bat.
27,722	Days, the average American's life expectancy.
46,399	Number of times the word *and* appears in the King James Bible.
76,115	ATMs (automatic teller machines) in use in the United States.
100,874	Pounds of food eaten by the average American in a lifetime.
116,000	People admired Matisse's *Le Bateau* at New York's Museum of Modern Art before it was noticed to be upside down.
252,055	People are added to the world's population every day.
343,000	Puerto Ricans take vitamins regularly.
451,000	Dollars paid for a rare baseball card in 1991.
507,635	Tons of broccoli eaten each year in the United States.
605,584	People visit Graceland, Elvis Presley's home, in an average year.
774,900	Square feet, the size of the Imperial Palace in Beijing, China, the world's largest palace.
861,500	American Indians live on reservations.
974,731	To 1, the odds against dying a nonviolent death, if you are a codfish.
1,000,000	Dots in the book.

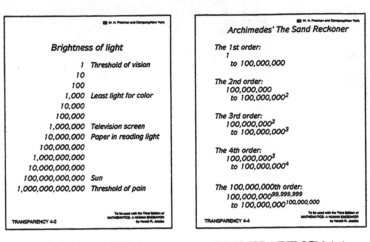

TRANSPARENCY 4-3   TRANSPARENCY 4-4

The eyes of a cat seem to glow at night. This is due to shiny fibers inside the cat's eyes that enable the cat to see in dim light. Even the range of light intensities that the human eye can see is enormous.   (Show Transparency 4-3.)

Brightness
of light

1	Threshold of vision
10	
100	
1,000	Least light for color vision.
10,000	
100,000	
1,000,000	Television screen
10,000,000	Paper in reading light
100,000,000	
1,000,000,000	
10,000,000,000	
100,000,000,000	Sun
1,000,000,000,000	Threshold of pain

In this table, the faintest light that a human

LESSON 2
Scientific Notation

eye adapted to the dark can see has a brightness of 1. The light needed to see color must be one thousand times as bright. What is the name of the number corresponding to the brightness of the sun? (One hundred billion.) What is the name of the number corresponding to light so bright that it is painful to look at? (One trillion.) How would the number corresponding to the brightness of a television screen be written as a power of ten? (10^6.) The number corresponding to the threshold of vision? (10^0.) The number corresponding to light 100 times brighter than the threshold of pain? (10^{14}.)

REVIEW OF LESSON 1

Transparency 4-4 is for use in discussing the Set II exercises of Lesson 1.

NEW LESSON

To introduce the new lesson, you might hold up a deck of playing cards. The number

TRANSPARENCY 4-5

of arrangements into which a deck of 52 playing cards can be shuffled is incredibly large.
An approximation of it is shown on Transparency 4-5. Could this number be written in a
simpler way? Because of the 8, it cannot be
written as a power of ten. Notice, however,
something that we might do. Draw a slash
through the 8, write a 1 above it, and 8 × in
front of it as shown below.

$$1$$
$$8 \times \cancel{8}0,000,000, \ldots$$

This change expresses the number as 8 times a
number that is a power of ten. What power of
ten is it? (10^{67}.) The approximate number of
ways in which a deck of playing cards can be
shuffled, then, can be written as

$$8 \times 10^{67}.$$

Explain that a number written in this way is
expressed in scientific notation:

$$a \times 10^b$$

in which a is a number that is at least 1 but less
than 10. Without this restriction on a, the
number could be written in other ways, such

as

$$80 \times 10^{66}.$$

It simplifies things to have exactly *one* way in
which to write a number in scientific notation.

You may also want to point out that when
a number in scientific notation is displayed on
a calculator, it does not appear in the form $a \times
10^b$, but rather as "$a$ b." The number 8×10^{67},
for example, appears as "8 67."

Further examples of how to change a number from decimal form into scientific notation
or from scientific notation into decimal form
will probably be needed before your students
begin the new lesson.

The following are appropriate.

Write in scientific notation.

1. 700,000.
2. 12,000,000,000.
3. 40×10^{20}.

Write in decimal form.

4. 5×10^4.
5. 9.3×10^7.

Number	Logarithm	Number	Logarithm
1	0	131,072	17
2	1	262,144	18
4	2	524,288	19
8	3	1,048,576	20
16	4	2,097,152	21
32	5	4,194,304	22
64	6	8,388,608	23
128	7	16,777,216	24
256	8	33,554,432	25
512	9	67,108,864	26
1,024	10	134,217,728	27
2,048	11	268,435,456	28
4,096	12	536,870,912	29
8,192	13	1,073,741,824	30
16,384	14	2,147,483,648	31
32,768	15	4,294,967,296	32
65,536	16		

TRANSPARENCY 4-6   TRANSPARENCY 4-7

LESSON 3

An Introduction to Logarithms

In his delightful book, *Odd Numbers**, Herbert McKay wrote:

At least one person in every newspaper office knows the following facts:
1 inch of rain weighs 100 tons per acre,
10 inches of snow = 1 inch of rain.

On the basis of this information, it is easy to arrive at some startling conclusions. Your students might enjoy some problems such as the following, but adapted to your school and city.

The school at which I teach has a campus of 35 acres. In an ordinary storm in which one inch of rain falls, the amount of water that falls on the campus is 3,500 tons, or, since there are 2,000 pounds per ton,

$$3,500 \times 2,000 = 7,000,000 \text{ pounds.}$$

Changing each number in a calculation such

as this to scientific notation makes a nice follow-up to the patterns for multiplying numbers in scientific notation:

$$(3.5 \times 10^3) \times (2 \times 10^3) = 7 \times 10^6.$$

Los Angeles, the city in which I live, has an area of approximately 300,000 acres. (This is based on an area of 464 square miles and the fact that 1 square mile = 640 acres.) One inch of rain falling on Los Angeles amounts to approximately

$$300,000 \times 100 = 30,000,000 \text{ tons of water!}$$

Writing each of these numbers in scientific notation gives

$$(3 \times 10^5) \times (1 \times 10^2) = 3 \times 10^7.$$

Transparencies 4-6 and 4-7 are for use in introducing the new lesson in accord with the text.

*Cambridge University Press, 1940.

LESSON 4

Decimal Logarithms

The following lesson plan includes both a review of Lesson 3 and an introduction to Lesson 4.

In 1971, the government of Nicaragua issued a set of postage stamps, one of which is pictured on Transparency 4-8. The subject which the stamps commemorated is named along the bottom of the stamp. You might ask if any of your students can translate the words ("the ten mathematical formulas that changed the face of the earth.") The stamp shown honors the invention of logarithms in the early seventeenth century by the Scottish mathematician John Napier.

The table below the stamp appeared in Napier's writings and is related to the table of logarithms of the last lesson.

I	II	III	IV	V	VI	VII
1	2	4	8	16	32	64	128

Draw a line through the line of Roman numerals and write the numbers they represent above them. Also label the two lines as shown here.

log	0	1	2	3	4	5	6	7
no.	1	2	4	8	16	32	64	128

What problem with the logarithms on the first line of this table corresponds to this multiplication problem with numbers from the second line of the table: $4 \times 8 = 32$? $(2 + 3 = 5.)$ To what kind of number sequence do the numbers in the second line belong? (The binary sequence, or a geometric sequence.) Recall that each of them can be written as a power of 2. (Write the powers below the numbers.)

no.	1	2	4	8	16	32	64	128
	2^0	2^1	2^2	2^3	2^4	2^5	2^6	2^7

Where do the logarithms of the numbers appear in these powers? (As the exponents.) The "magic" of the table comes from the fact that the logarithms are exponents:

$$4 \times 8 = 32$$

can be written as

Binary logarithms		Decimal logarithms		
Logs	Numbers	Logs	Numbers	
0	$2^0 = 1$	0	$10^0 =$	1
1	$2^1 = 2$	1	$10^1 =$	10
2	$2^2 = 4$	2	$10^2 =$	100
3	$2^3 = 8$	3	$10^3 =$	1,000
4	$2^4 = 16$	4	$10^4 =$	10,000
5	$2^5 = 32$	5	$10^5 =$	100,000
6	$2^6 = 64$	6	$10^6 =$	1,000,000

TRANSPARENCY 4-9

$$2^2 \times 2^3 = 2^5,$$

which means

$$\underbrace{2 \times 2}_{2} \times \underbrace{2 \times 2 \times 2}_{3} = \underbrace{2 \times 2 \times 2 \times 2 \times 2}_{5}$$

What if 2, called the *base* in these equations, is replaced with a different number? For example, is it true that

$$10^2 \times 10^3 = 10^5?$$

Yes, because

$$\underbrace{10 \times 10}_{2} \times \underbrace{10 \times 10 \times 10}_{3} = \underbrace{10 \times 10 \times 10 \times 10 \times 10}_{5}$$

This suggests that logarithms do not have to be based on the number 2. They can be based on another number, such as 10, instead.

Look again at the postage stamp. The words "ley de Napier" refer to "the law of Napier." Logarithms are exponents and in the formula

$$e^{\ln N} = N$$

the symbol $\ln N$ represents a logarithm (ln) of a number (N). Notice that it appears in the formula as an *exponent*.

Show Transparency 4-9. These tables show the difference between binary logarithms, based on the number 2, and decimal logarithms, based on the number 10.

Binary logarithms

logs	0	1	2	3	4	5	6
numbers	2^0	2^1	2^2	2^3	2^4	2^5	2^6

Decimal logarithms

logs	0	1	2	3	4	5	6
numbers	10^0	10^1	10^2	10^3	10^4	10^5	10^6

Decimal logarithms are used more frequently than binary logarithms because their connection to the decimal system of writing numbers is so simple. For example, which is easier to remember: that the binary logarithm of 64 is 6, or that the decimal logarithm of 1,000,000 is 6?

It is for this reason that when the button labeled "log" on a scientific calculator is pushed, it gives the *decimal* logarithm of the number.

If you have an overhead scientific calculator, at this point you might construct a two-place table of decimal logs as shown below:

logs	0	0.30	0.48		0.70
numbers	1	2	3	4	5

logs		0.85		0.95	1
numbers	6	7	8	9	10

(Point out that although this is an easy way to find logarithms, they were figured out by other methods long before the invention of the calculator.)

Can you figure out the three missing logarithms?

$$(2 \times 2 = 4, \text{ so } 0.30 + 0.30 = 0.60;$$
$$2 \times 3 = 6, \text{ so } 0.30 + 0.48 = 0.78;$$
$$2 \times 4 = 8, \text{ so } 0.30 + 0.60 = 0.90.)*$$

---

*By this reasoning, someone might conclude that $\log 9 = 0.96$ since $3 \times 3 = 9$ and $0.48 + 0.48 = 0.96$. If a student asks why the table gives 0.95 instead, explain that when we write these numbers to 2 decimal places, they are not exact. An extra digit, for example, reveals that $\log 3 = 0.477$; $0.477 + 0.477 = 0.954$ and, to *three* decimal places, $\log 9 = 0.954$.

TRANSPARENCY 4-10

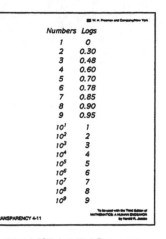

Numbers	Logs
1	0
2	0.30
3	0.48
4	0.60
5	0.70
6	0.78
7	0.85
8	0.90
9	0.95
10^1	1
10^2	2
10^3	3
10^4	4
10^5	5
10^6	6
10^7	7
10^8	8
10^9	9

TRANSPARENCY 4-11

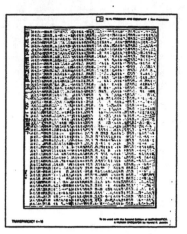

TRANSPARENCY 4-12

LESSON 5
Logarithms and Scientific Notation

Show Transparency 4-10. Howard Eves, in his book *An Introduction to the History of Mathematics* (Saunders, 1990), tells of an amusing incident in John Napier's life. Napier owned a large estate near Edinburgh, Scotland, on which he raised grain. One of his neighbors raised pigeons and the birds would fly over to his land to eat the grain. Napier threatened to catch the pigeons if the neighbor did not take measures to keep them off his property. The neighbor told Napier that he was welcome to keep any birds that he could capture. If you have ever tried to catch a pigeon, you know that, although it may look easy, it is actually almost impossible. So the neighbor thought he was perfectly safe in making the offer. The next day, however, he was astonished to see his pigeons staggering and reeling on Napier's lawn. Napier had soaked peas in brandy and had scattered them about for the pigeons to eat. He had no trouble at all in catching every one of them.

REVIEW OF LESSON 4

Transparency 4-11 may be useful in discussing some of the exercises of Lesson 4.

INTRODUCTION TO NEW LESSON

Transparencies 4-12 and 4-13 are for possible use in presenting the new lesson in accord with the text. The Chinese logarithm table from which the page pictured on Transparency 4-12 is taken filled a book of more than 600 pages. If you have access to a copy of a book of logarithms, you might pass it around the class. Copies of such books are available in some used book stores. The one I use is *A New Manual of Logarithms to Seven Places of Decimals*, edited by Dr. Bruhns and published by the Charles T. Powner Co. in Chicago in 1967. This is undoubtedly one of the last such books to be published before the advent of the elec-

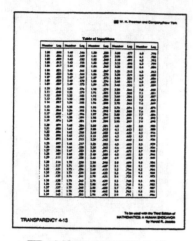

TRANSPARENCY 4-13

tronic calculator in 1972 rendered them totally obsolete.

The table on page 218 of the textbook (and on Transparency 4-13) is adequate for all of the exercises in this and the following lessons.

Some examples such as the following may be helpful.

Find the logarithms of the following numbers.

1. 2.15   (0.332)
2. 2.15×10^7 (7.332)
3. 21,500   (4.332)

Find the numbers having the following logarithms.

4. .806   (6.4)
5. 13.806   (6.4×10^{13})
6. 2.806   (640)

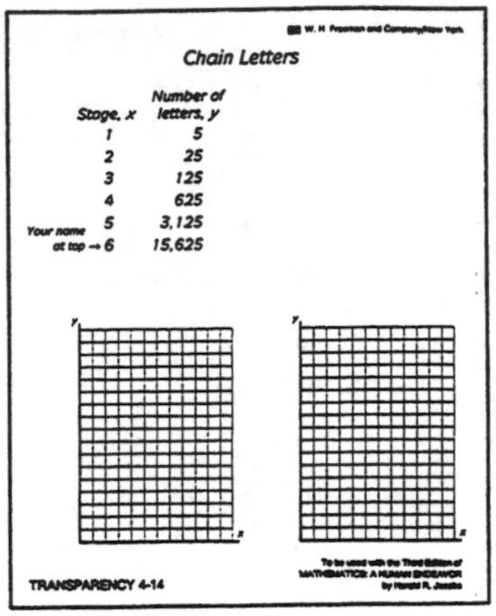

TRANSPARENCY 4-14

Exponential Functions

More on chain letters

After reminding your students of the chain letter craze that started in Denver in 1935, hand out duplicated copies of Transparency 4-14. This list shows the numbers of letters in successive stages of the chain, starting with the 5 letters you send out.

Stage, x	Number of letters, y
1	5
2	25
3	125
4	625
5	3,125
Your name at top → 6	15,625

If no one breaks the chain, this list shows that when your name appears at the top of the list, you will get 15,625 dimes, or $1,562.50.

Have your students express the y numbers as products and powers of 5 as shown here:

x	y	
1	$5 = 5$	$= 5^1$
2	$25 = 5 \times 5$	$= 5^2$
3	$125 = 5 \times 5 \times 5$	$= 5^3$
4	$625 = 5 \times 5 \times 5 \times 5$	$- 5^4$
5	$3,125 = 5 \times 5 \times 5 \times 5 \times 5$	$= 5^5$
6	$15,625 = 5 \times 5 \times 5 \times 5 \times 5 \times 5$	$= 5^6$

It is easy to discover a formula for this function by noticing that each of the y numbers is a power of 5:

$$y = 5^x.$$

Such a function is called *exponential* because one of the variables, x, appears as an exponent.

What would the graph of this function look like? Notice that on the first grid there is plenty of room for the x-numbers. We can even space them 2 units apart. The last y-number, on the other hand, is so big that to fit it on the graph, we have to let 1 unit on the y-axis represent 1,000.

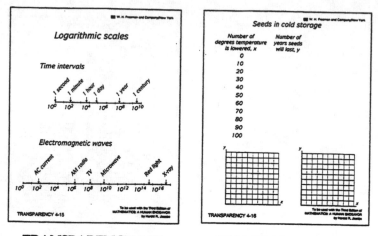

TRANSPARENCY 4-15   TRANSPARENCY 4-16

After your students have discovered that the graph is hard to draw because the first few values of *y* are so small in comparison to the last value, remark that in such a case it is sometimes useful to use a logarithmic scale instead. To see why, add a third column to the list on the transparency, in which the *y*-values are represented by their logarithms. (Your students will need some help in doing this They can use the table on page 218 and round each log to the nearest tenth.)

x	*log y*
1	0.7
2	1.4
3	2.1
4	2.8
5	3.5
6	4.2

Graph the resulting data on the second grid, letting 4 units on the *log y*-axis represent 1. Notice that not only are the logs of the *y*-numbers easier to graph, but also that the graph is a straight line. This happens whenever an exponential function is graphed in this way. A graph of it on axes with ordinary scales is a curve, but a graph of it in which a logarithmic scale is chosen for *y* is a straight line.

REVIEW OF LESSON 5

The scales on Transparency 4-15 may be useful in discussing some of the Set II exercises of Lesson 5.

ALTERNATE LESSON

The following example is similar to the chain letter example but has the disadvantage of not having a simple formula to make the meaning of exponential function clear.

Some seeds thought to be 10,000 years old were found in frozen ground in Canada in 1954. After being thawed out, they were eventually planted and germinated, producing some plants called Arctic lupine.

It has been claimed that when seeds are kept in cold storage, their life span is doubled for each 10 degrees Fahrenheit that their temperature is lowered. Hand out duplicated copies of Transparency 4-16. Suppose that this claim is true and that some seeds stored at 75°F will last 10 years.

Copy and complete the first table on the
basis of this information.

Number of degrees temperature is lowered, x	Number of years seeds will last, y
0	(10)
10	(20)
20	(40)
30	(80)
40	(160)
50	(320)
60	(640)
70	(1,280)
80	(2,560)
90	(5,120)
100	(10,240)

After your students have graphed the function on the first grid, help them complete the
second table.

Number of degrees temperature is lowered, x	Log of number of years seeds will last, y
0	(1)
10	(1.3)
20	(1.6)
30	(1.9)
40	(2.2)
50	(2.5)
60	(2.8)
70	(3.1)
80	(3.4)
90	(3.7)
100	(4.0)

The results can be graphed on the second grid.

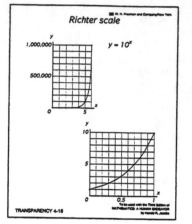

Powers of Ten	
10^{25} The universe	10^4
10^{24} Our cluster of	10^3
10^{23} galaxies	10^2
10^{22}	10^1
10^{21} Our Milky	10^0 A human
10^{20} Way galaxy	10^{-1} being
10^{19}	10^{-2}
10^{18}	10^{-3}
10^{17}	10^{-4}
10^{16}	10^{-5} A blood cell
10^{15}	10^{-6}
10^{14}	10^{-7} DNA
10^{13} Our solar	10^{-8}
10^{12} system	10^{-9}
10^{11}	10^{-10} The atom
10^{10}	10^{-11}
10^9 The orbit of	10^{-12}
10^8 the moon	10^{-13}
10^7 The earth	10^{-14} The atomic
10^6	10^{-15} nucleus
10^5	10^{-16} Quarks

TRANSPARENCY 4-17   TRANSPARENCY 4-18   TRANSPARENCY 4-19

Review of Chapter 4

First day

Having become acquainted with logarithmic scales, your students would enjoy seeing the film *Powers of Ten*. It was made in 1978 by the office of Charles Eames and is available on video tape and laser disc. The film begins with a view of a picnic in a park in Chicago and takes the viewer outward to the edge of the universe and then inward to the nucleus of an atom—all in just nine minutes. It is a remake of an earlier film titled *A Rough Sketch for a Proposed Film Dealing with the Powers of Ten and the Relative Size of Things in the Universe* and was inspired by the book *Cosmic View: The Universe in 40 Jumps* by Kees Boeke (The John Day Company, 1957).

A book version of the film is *Powers of Ten* by Philip and Phylis Morrison and The Office of Charles and Ray Eames (Scientific American Library, 1982). Another book based on the same concept is *From Quark to Quasar—Notes on the Scale of the Universe* by Peter Cadogan (Cambridge University Press, 1985).

Transparency 4-17 lists some of the major stages of the film.

REVIEW OF LESSON 6

Transparencies 4-18, 4-19, 4-20, and 4-21 are for use in discussing the exercises of Lesson 6.

Second day

Show Transparency 4-22. The label on a bottle of a certain brand of shampoo states that it has a "pH 5" formula. The pH scale was invented by a Danish chemist to measure how acidic or alkaline a solution is. Show Transparency 4-23. The pH of a solution can range from 0 to 14, with 0 the most acidic, 7 neutral, and 14 the most alkaline.

According to where they appear on this scale, which is more acidic: black coffee or beer? (Beer.) Which is more acidic: soap or water? (Water.)

Each "hydrogen ion concentration" in the table can be expressed as a power of 10. How

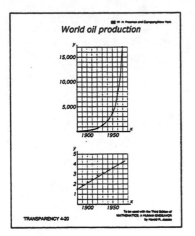

TRANSPARENCY 4-20

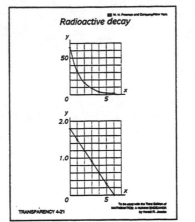

TRANSPARENCY 4-21

TRANSPARENCY 4-22

can the number 1 be expressed as a power of 10? (10^0.) The numbers 0.1 and 0.01 can be expressed as 10^{-1} and 10^{-2} respectively. (Write these three powers on the table.)

How is the exponent of each 10 related to the pH? (It is the opposite, or negative, of it.) How can the hydrogen ion concentration of drain cleaner be expressed as a power of 10? (10^{-14}.)

If you have "acid indigestion," your stomach is too acidic. To neutralize the acid, you might make it more alkaline by taking a medicine such as Alka-Seltzer. What does this do to the hydrogen ion concentration in your stomach? (It decreases it.) What does it do to the pH? (It increases it.)

Recall that when a number is expressed as a power of 10, the exponent of 10 is the decimal logarithm of the number. What is the decimal logarithm of 10^{-14}? (-14.) The letters p and H in the name pH refer to "power" and "hydrogen" respectively. The pH scale is used in such fields as chemistry, biology and medicine. To interpret it correctly, scientists and physicians have to understand decimal logarithms.

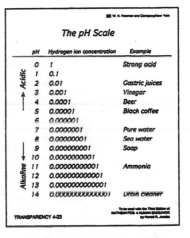

TRANSPARENCY 4-23

CHAPTER **5**

Symmetry and Regular Figures

This chapter is an informal excursion into geometry. Centered on regular polygons, it includes material not usually found in geometry courses. The approach is primarily intuitive and descriptive, the students being given the opportunity to make interesting discoveries on their own.

The first lesson presents the concepts of line and rotational symmetry and includes a simple yet highly intriguing experiment with mirrors. Regular polygons are introduced in the second lesson. Included in this lesson are exercises in which the students construct regular polygons by inscribing them in circles and draw regular polygons by using a circular protractor.

Relations between regular polygons in the plane—the regular and semiregular tessellations, called "mathematical mosaics" in the text—are considered in the third lesson. The remaining three lessons deal with relations between regular polygons in space—the regular polyhedra, the Archimedean semiregular polyhedra, and regular pyramids and prisms. After discovering Euler's formula, the students prove that it holds for several types of geometric solids. Several experiments are included in which the students make drawings and build models. Artistic students, as well as those who simply like to work with their hands, especially enjoy this part of the course. They should be encouraged to explore freely and to look for pleasing new constructions. They may well discover something new (for them if not for us).

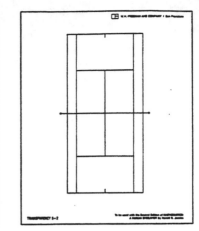

TRANSPARENCY 5-1   TRANSPARENCY 5-2

First day

Show Transparency 5-1. This figure is used in a famous psychological test named after its inventor, Hermann Rorschach. It is an ink blot and the person being tested is asked to describe what he or she sees in it. What do you see?

Each of the figures used in the Rorschach test is an ink blot and, since the test was first given in 1921, it has become famous throughout the world. In his book *Reality's Mirror: Exploring The Mathematics of Symmetry* (Wiley, 1989), Bryan Bunch wrote:

What characteristics do the Rorschach blots have that make them so effective? Perhaps it is a direct result of the process by which they were made.

The result of folding the paper over an ink blot is to produce a shape in which the left side is exactly like the right one. Most animals also share this characteristic ... For this reason, the Rorschach blots are very often perceived as animals ... The [person's] answer is a response to symmetry, not to a blob.

It is fairly easy to guess where the paper was folded to produce this ink blot. Draw the line on the transparency and call attention to some of the matching features on the two sides of the line. Also point out that a mirror could be placed on this line to reflect either half of the figure so that it reproduces the other half and explain that such a figure is said to have "line symmetry."

A figure can have more than one line of symmetry. Show Transparency 5-2. This is a scale drawing of the floor of a tennis court. Where could a mirror be placed on this drawing so that half of the figure would be reflected to produce the other half? (Along either a vertical line through its center or the horizontal line containing the net.) The figure, then, has two lines of symmetry.

It also has another kind of symmetry. (Rotate the transparency 180°.) Rotating the drawing 180° does not change its appearance. When

75

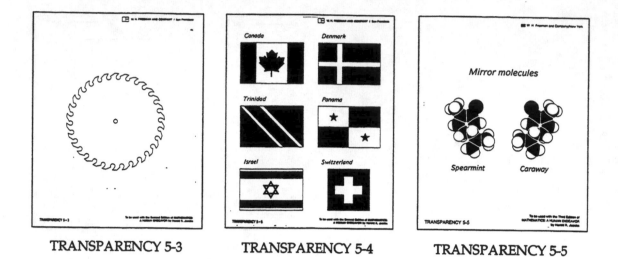

TRANSPARENCY 5-3 TRANSPARENCY 5-4 TRANSPARENCY 5-5

two tennis players change ends, their viewpoint of the court remains unchanged. Because of this, the floor of a tennis court is said to have "rotational symmetry." It can be rotated through an angle of less than 360° so that it coincides with its original position.

A figure can have rotational symmetry without having line symmetry. Show Transparency 5-3. (The most effective way to use the transparency is to make an extra copy of the figure on heavy-weight film. It can be attached to the transparency by putting a straight pin through the centers of the figures and then cutting off the rest of the pin.) This figure shows a blade of a circular saw. The blade does not have line symmetry because all of its teeth go in one direction around its rim. A mirror would reverse the direction of the teeth. The figure does have rotational symmetry, however, because it can be rotated through many different angles so that it coincides with its original position. (Turn the overlay to illustrate some of them.) How could you determine the measures of some of these angles? (Count the number of teeth; there are 30. Divide 360° by 30 to find the smallest angle: 12°. All of the angles are multiples of 12°.) Because the blade has 30 teeth, there are 30 different angles, up to and including 360°. The figure is said to have "30-fold" rotational symmetry.

Show Transparency 5-4. Many flags have some type of symmetry. What symmetries do these flags have? (Canada: one vertical line of symmetry. Denmark: one horizontal line of symmetry. Trinidad: 2-fold rotational symmetry. Panama: no symmetry. Israel: two lines of symmetry, one vertical and one horizontal, and 2-fold rotational symmetry. Switzerland: four lines of symmetry and 4-fold rotational symmetry.)

With this introduction to symmetry, the students are prepared to read Lesson 1 and begin the exercises. Note that a small square of tracing paper (3-by-3 inch) is needed for exercise 20.

Second day

Show Transparency 5-5. Chemists have made a surprising discovery about molecules that are mirror images of each other.* An example

_____

*_Molecules,_ by P. W. Atkins (Scientific American Library, 1987.)

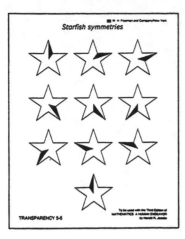

TRANSPARENCY 5-6

is the compound carvone. The "left-handed" version of carvone is the main ingredient of spearmint, a popular flavor used in chewing gum. The "right-handed" version of carvone is responsible for the odor of caraway seeds, often used on crackers. Amazingly, except for the fact that the molecules of these two flavors are mirror images of each other, they are identical in every other way: both contain 10 carbon atoms, 18 hydrogen atoms, and 1 oxygen atom connected in the same way.

Chemical companies are now producing mirror image molecules such as these for different purposes. The Eli Lilly Company makes a drug called Darvon which is used as a sedative. They also make its mirror-image twin, Novrad, which is used as a cough medicine. Perhaps most surprising of all is that in some cases, one version of a molecule can be capable of curing disease whereas its mirror image may be poisonous.

Transparency 5-6 is for possible use in discussing the Set I exercises on the symmetries of the starfish.

Two mirrors hinged together are needed to do the Set II exercises. Mirrors that are 4-by-6 inches are available from Creative Publica-tions, 5040 West 111th Street, Oak Lawn, IL 60453, phone 800-624-0822.

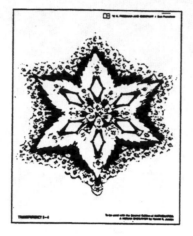

TRANSPARENCY 5-7　　　　TRANSPARENCY 5-8

LESSON 2
Regular Polygons

First day

Show Transparency 5-7.* Many early Hollywood musicals featured highly symmetric patterns of dancers. This drawing shows one of these patterns, which almost looks as if it could have been produced in a kaleidoscope. Created by Busby Berkeley for the 1934 Warner Brothers film, *Dames*, it consists of 78 dancers being photographed by an overhead camera.

How many lines of symmetry does the pattern seem to have? (After your students have had time to discover all six, draw them on the transparency. Even though the figure is not perfectly symmetric, draw all six lines so that they pass through the center of the figure.)

Because of the way in which the people are

---

*If you have sets of hinged mirrors available, this is more effective if you hand out duplicated copies of the transparency together with rulers and mirrors so that your students can directly participate.

positioned in the figure, it does not seem as if it could have been produced by mirrors. Why not? (The mirrors would have to be placed on symmetry lines, which cut some of the people in half.)

At what angles can you set the mirrors on the figure so that what they reflect reproduces the entire pattern? (Any pair of lines will do, so the mirrors can be placed at angles of 30°, 60°, 90°, 120°, 150°, and 180°.)

Does the pattern seem to have rotational symmetry? (Yes.) What is the smallest angle through which it could be rotated to coincide with its original position? (60°.)

REVIEW OF LESSON 1

Transparency 5-8 is for use in discussing some of the Set II exercises of Lesson 1.

NEW LESSON

To introduce the new lesson, you might hand out duplicated copies of Transparency

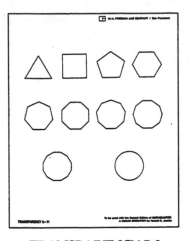

TRANSPARENCY 5-9

5-9 on which your students can note the names and symmetries of the regular polygons. Because most students will not have their own compasses, you may want to hand out rulers and compasses so that the constructions can be done in class. If you can afford the time, you might want to lead your students through a series of regular polygon constructions in class, possibly including those with 3, 4, 5, 6, and 8 sides. (A method for constructing a regular pentagon is outlined on page 320 of the text.)

Students who have had little practice in using a compass often have trouble with it slipping or tearing the paper. This is generally caused by working on a single sheet of paper on a hard desk top and can be prevented by putting several extra sheets of paper underneath to serve as padding for holding the compass point.

I recommend starting with circles with a radius of at least an inch; some of the constructions require working in the regions outside the circles, and so the circles should have ample space around them.

Note that no attempt is made to prove that the methods used in the constructions are correct or even to explain to the student why they work. Most students tend to accept the

produces without question and are perfectly satisfied with merely completing the drawings. The proofs appropriate for a course in geometry are beyond the scope of this one. Although individual students should be encouraged to ask why these methods work, I urge caution against introducing theoretical explanations that your students may not consider necessary or that they may not yet be able to comprehend.

Determining the corners of a square using only a compass (exercise 13 of Set I), is a relatively straightforward example of a Mascheroni construction. The demonstration that it is correct makes an interesting exercise for good geometry students. If we let r represent the length of the radius of the circle below,

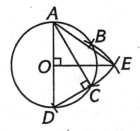

then the length of the side of the required square is readily shown to be $r\sqrt{2}$.

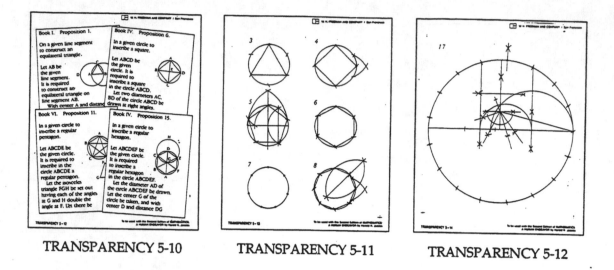

TRANSPARENCY 5-10   TRANSPARENCY 5-11   TRANSPARENCY 5-12

Because triangle ACD is a right triangle with leg CD = r and hypotenuse AD = $2r$, leg AC = $r\sqrt{3}$. But AC = AE. So in right triangle AOE, we have leg AO = r and hypotenuse AE = $r\sqrt{3}$. Hence the other leg OE = $r\sqrt{2}$, the length needed for the side of the square.

The eighteenth-century Italian mathematician Lorenzo Mascheroni proved that any construction that can be done with a straightedge and compass can also be done with a compass alone. See pages 147-152 of *What Is Mathematics?* by Richard Courant and Herbert Robbins (Oxford University Press, 1941) for a brief discussion of Mascheroni constructions.

Second day

Show Transparency 5-10. The methods used in yesterday's lesson to construct regular polygons have been known for a long time. These pages are translations from Euclid's "Elements," written in the third century B.C. In this book, Euclid explained how to construct regular polygons having three, four, five and six sides. He did not know a method for constructing a regular polygon with seven sides and, in the more than two thousand years since Euclid's time, such a method has never been found.

Show Transparency 5-11. The problem of constructing a regular polygon with a given number of sides is equivalent to being able to locate that number of equally spaced points on a circle. (Add the overlay.) Some polygons, such as the regular hexagon, are easy to construct. Others, such as the regular pentagon, are considerably more difficult. Some, such as the regular heptagon, are impossible! This was proved in 1796 by Carl Friedrich Gauss when he was nineteen years old; Gauss showed that seven equally spaced points cannot be located on a circle with just a straightedge and compass.

Show Transparency 5-12. At the time Gauss proved this, he made a surprising discovery: it *is* possible to use these tools to locate 17 evenly spaced points on a circle. The method for doing so, as you might imagine, is quite complicated. It makes the constructions that we have done seem simple in comparison. (Add the overlay.) Through a long series of steps, the vertical segments at the left and right are drawn. These segments cut off an arc of the circle that is exactly two-seventeenths of it. This arc is then divided in half to produce

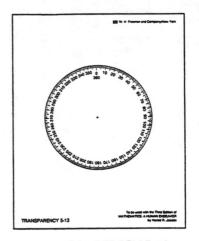

TRANSPARENCY 5-13

an arc that is exactly one-seventeenth of the
circle. The method is described, together with
a brief history of the constructability of the
regular polygons, on pages 26-28 of *Introduction to Geometry*, second edition, by H. S. M.
Coxeter (John Wiley and Sons, 1969.)

Each student will need a sheet of tracing
paper for the Set II exercises of Lesson 2. The
circular protractor is reproduced on Transparency 5-13 for possible use in class.

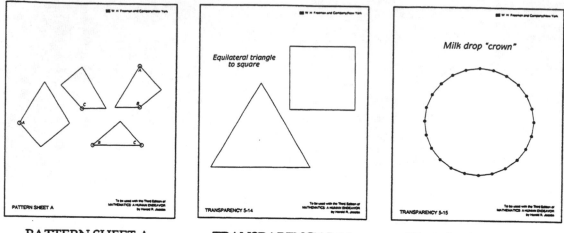

PATTERN SHEET A   TRANSPARENCY 5-14   TRANSPARENCY 5-15

LESSON 3
Mathematical Mosaics

First day

The classic book *Mathematical Snapshots* by H. Steinhaus (Oxford University Press, 1969), opens with an illustration of a clever model consisting of four pieces hinged at their corners to form an open chain. If the chain is swung around in one direction, the four pieces form a square; if it is swung around in the opposite direction, the pieces form an equilateral triangle. The patterns on Pattern Sheet A can be used to make a copy of this hinged model that you can show on your overhead projector. Copy the four pieces onto file cards and then use a pin to poke holes into the cards at the indicated hinge points. Cut the pieces out and use three straight pins to connect them at the hinge points (A to A, B to B, and C to C.) The model is now ready to use.

Begin by showing Transparency 5-14, with the chain of the four pieces placed below the two figures. It is possible to take any regular polygon and cut it into pieces that can be rearranged to form any other regular polygon.

Such a puzzle is called a dissection puzzle and the four pieces shown here are a solution for rearranging an equilateral triangle to form a square. Show both arrangements of the pieces.

After doing this, you might ask a student who has solved the Set III exercise of Lesson 2 to demonstrate the solution on the projector.

Transparencies 5-15, 5-16, and 5-17 are for possible use in discussing the Set II exercises of Lesson 2.

NEW LESSON

A good way to introduce the new lesson is to show your students an actual honeycomb. The bee is an extraordinary architect, for the design of the honeycomb includes much advanced mathematics. If you look closely at a honeycomb, the first thing you will probably notice is that, not only are all its cells exactly alike, but they have the shape of regular hexagons. Show Transparency 5-18. Although this design is seen in such man-made objects as bathroom tiles and chicken wire, bees were

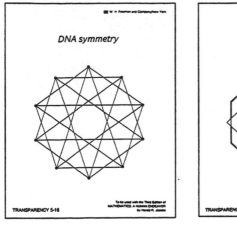

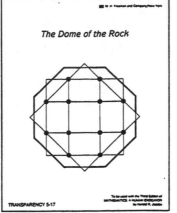

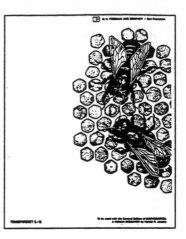

TRANSPARENCY 5-16   TRANSPARENCY 5-17   TRANSPARENCY 5-18

using it long before humans ever thought of these things. Use the transparency to introduce the notation 6-6-6.

Transparency 5-19 with its polygons (Pattern Sheet B) and overlay can now be used to present the rest of the lesson in accord with the text. The polygons are especially effective if cut from heavyweight plastic with each kind a different color. After the point on the transparency has been surrounded with two octagons and a square as illustrated on page 267 of the text, the overlay can be added to show how the pattern can be extended to fill the plane.

To do the Set I exercises of the new lesson, each student will need the following materials:

One 9-by-12-inch sheet of oak tagboard

Tracing paper (or a duplicated copy of Pattern Sheet C)*

---

*The directions say to use tracing paper to copy the figure on page 269. I recommend, however, that you duplicate copies of the figure (reproduced on Pattern Sheet C) to hand out to your students. This will both save time and result in more accurate polygons.

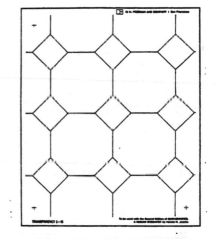

TRANSPARENCY 5-19

A small manila envelope for storing the finished set of polygons

Scissors

A compass

A ruler

The first time that the experiment is done, extra time will be needed to prepare the set of polygons. If enough cardboard is available and you have two classes, you might have

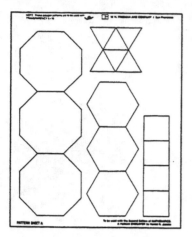

PATTERN SHEET B

PATTERN SHEET C

Cellular telephones

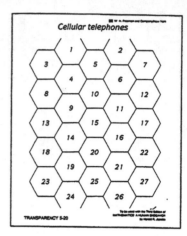

TRANSPARENCY 5-20

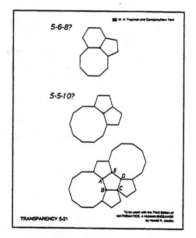

TRANSPARENCY 5-21

both classes prepare the materials so that those that have been badly made can be discarded; this will leave an ample set of materials for even very large classes that you might have in the future. To keep the amount of wasted material to a minimum and insure that all of the polygons are cut from one piece of tagboard, they should be cut out in the order listed on page 270; that is, from largest to smallest.

Emphasize that in this experiment we are trying to discover combinations of regular polygons that will surround a *single point*. Some students will get carried away and attempt to form extended mosaics; they can be quickly put on the right track if you move about the room and observe what everyone is doing.

As the students will learn from the Set II exercises, the problem of surrounding a single point is not equivalent to determining a complete mosaic because some of the combinations that satisfy the first condition cannot be extended to produce the second.

Second day

Show Transparency 5-20. Cellular telephones have become popular in large cities. In their book *Signals, The Science of Telecommunications* (Scientific American Library, 1990), John R. Pierce and A. Michael Noll explain the basis for the word "cellular":

In cellular telephony, the metropolitan area is divided into a number of clusters, each cluster consisting of seven cells. Each cell is served by its own low-powered radio transmitter, called a base station, and is allocated

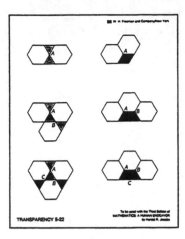

TRANSPARENCY 5-22

about 47 two-watt radio channels. A seven-cell cluster serves a total of 333 channels. . .

When a user leaves the jurisdiction of one cell, the user must be switched to one of the channels serving an adjacent cell. . . A typical cell has a radius of from 6 to 12 miles. . .

Portable cellular telephones will continue to become smaller and less expensive. Dick Tracy's wrist telephone might indeed be closer than we think.

So cellular telephones get their name from the same arrangement of cells that bees use in building their honeycombs!

Transparencies 5-21 and 5-22 are for use in discussing the Set II exercises of Lesson 3. Students who have done the Set III exercise on Escher's horseman will be interested in looking at the definitive book on Escher's mosaics: *Visions of Symmetry,* by Doris Schattschneider (W. H. Freeman and Company, 1990).

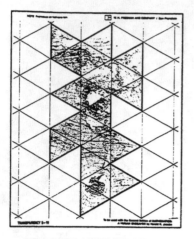

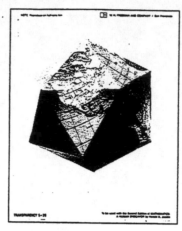

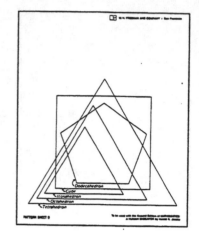

TRANSPARENCY 5-23   TRANSPARENCY 5-24   PATTERN SHEET D

LESSON 4

Regular Polyhedra: The Platonic Solids

First day

Show Transparency 5-23. This map of the earth was created by Buckminster Fuller, the inventor of the geodesic dome. It consists of twenty equilateral triangles, each of which contains an equal amount of the earth's surface. Notice that the triangles are part of the mosaic in which six equilateral triangles surround each corner point. If the map is cut out of the mosaic and folded along the sides of the triangles, it can be made into a three-dimensional solid called an icosahedron. A punch-out model of this solid, called the Spaceship Earth Dymaxion Globe, can be obtained from some map stores and from The Buckminster Fuller Institute, 1743 South La Cienega, Los Angeles, CA 90035, phone (213) 837-7710.

Show either Transparency 5-24 or a model of the map in its three-dimensional form. The orientation of the earth's continents on the map's surface has been arranged so that the corners of the icosahedron, which are the regions of greatest distortion, fall either on oceans or on shore lines.

In a way, this solid is like a mosaic because it is an arrangement of regular polygons in which the polygons share their sides and the polygons at each corner are alike in number, kind, and order. The triangles are its faces, their sides are its edges, and their corners are its corners. The solid is an example of a regular polyhedron, the subject of today's lesson.

To acquaint your students with the regular polyhedra, you might show models of them to your class. The photographs on pages 277-278 of the text are of models that I constructed from posterboard and tape. Patterns for their faces are on Pattern Sheet D of the book of transparency masters. (The edges vary in length so that all five models are about the same size.)

In place of, or in addition to, showing models of the polyhedra, you might want to hand out duplicated copies of Transparency 5-25, on which your students can make some brief notes.

Note the materials needed for making the

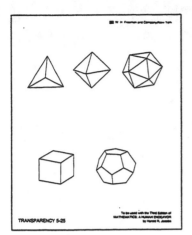

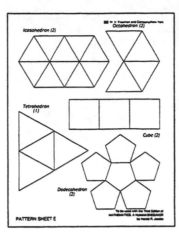

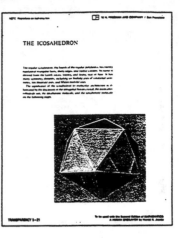

TRANSPARENCY 5-25   PATTERN SHEET E   TRANSPARENCY 5-26

models of the regular polyhedra described in the Set I exercises. If your students make these models in class, each person will need:

Tracing paper (or a copy of Pattern Sheet E)

Seven 4-by-6 inch unlined file cards

A ruler

A compass (or a straight pin)

Scissors

Tape

If the "pop up" version of the dodecahedron is made, each student will need:

Two 4-by-6 inch pieces of stiff card (such as a file folder) in place of two of the file cards mentioned above

A size 19 rubber band

I have discovered that supposedly identical rubber bands manufactured by different companies seem to vary considerably in length and stretch. If the bands used for this model are too tight, it will not take a very satisfactory shape as it pops up. Eberhard Faber "Star" rubber bands, size 19, work especially well. A

one-ounce box contains approximately 150 rubber bands. Some students will need to be shown how to weave the band around the corners of the two bowls.

Second day

Most classes will probably need two days to complete the Set I exercises if the work is done in class. At the beginning of the second day, you might show your class Transparencies 5-26 and 5-27, taken from *The Architecture of Molecules* by Linus Pauling and Roger Hayward (W. H. Freeman and Company, 1964). Point out that all five regular polyhedra are discussed in the book because a knowledge of these solids is important in understanding the arrangements of atoms in molecules and crystals. The boron crystal pictured on Transparency 5-27 is nearly as hard as diamond. Its atoms are arranged at the corners of icosahedra linking together by other atoms.

Third day

Many viruses, including the ones that cause

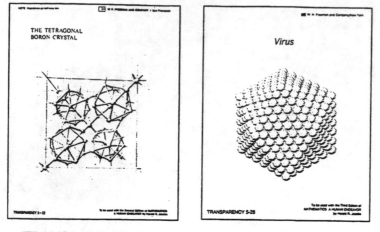

TRANSPARENCY 5-27   TRANSPARENCY 5-28

measles, herpes, and AIDS, have been discovered to have a core in the shape shown on Transparency 5-28.* You have made a model of this shape. What is it? (The icosahedron.) You used your model to count its numbers of faces, corners and edges. These numbers are related to the three-dimensional symmetries of the icosahedron, which seem to relate to why many viruses have this particular shape.

As an introduction to some of the exercises of Set II, guide your students in deducing the numbers of corners and edges of the icosahedron from the following facts:

An icosahedron has 20 triangular faces.
Five triangles meet at each corner.
Two triangles meet at each edge.

The set III exercise, based on the Back in the Box puzzle, affords a pleasing example of break-through thinking in solving problems. Materials needed for each student are:

Two 6-by-6 inch squares of oak tagboard

Tape

A ruler

A compass or a protractor (for drawing the equilateral triangles)

Scissors

---

*An especially good electron micrograph revealing the icosahedral shape of a virus appears in *Viruses,* by Arnold J. Levine (Scientific American Library, 1992), page 18.

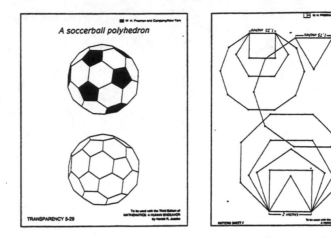

TRANSPARENCY 5-29   PATTERN SHEET F

Semiregular Polyhedra

First day

A soccerball is a good visual aid for introducing the lesson on the semiregular polyhedra. Show a soccerball* and Transparency 5-29. The pattern on the ball makes it look very much like a polyhedron. What regular polygons seem to appear in it? Call attention to the fact that each corner is surrounded by one pentagon and two hexagons. Circle several corners on the second figure, writing the numbers 5, 6, and 6 in the appropriate places.

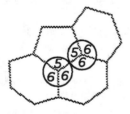

---

*You may want to mark the grooves between the edges of the hexagons with a black felt-tip marker so that the individual faces can be more easily seen.

This leads easily to the definition of "semiregular polyhedron," following which your students are ready to read the lesson and begin the exercises.

A set of the thirteen Archimedean solids make an attractive addition to a mathematics classroom. The photographs on pages 290 and 291 of the text are of models that I made from posterboard and tape. The models have a uniform edge length of 2 inches except for the truncated dodecahedron, which has an edge length of $1^3/_4$ inches, and the great rhomb-icosidodecahedron, which has an edge length of $1^1/_4$ inches. I used the following color scheme: triangle, silver; square, purple; pentagon, light yellow; hexagon, gold; octagon, fluorescent blue; decagon, white. Tracing paper templates made from Pattern Sheet F are useful in copying the polygons onto the posterboard and a paper cutter can be used for most of the cutting. Detailed information about each solid is given in *Mathematical Models* by H. Martyn Cundy and A. P. Rollett (Oxford University Press, 1961), pages 100-

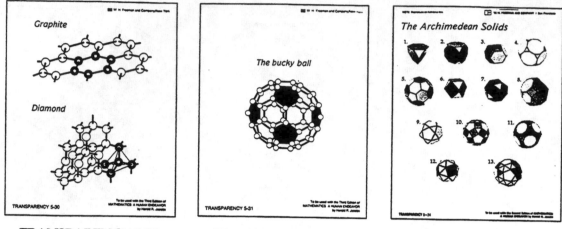

TRANSPARENCY 5-30   TRANSPARENCY 5-31   TRANSPARENCY 5-32

115. Also recommended is the book *Polyhedron Models* by Magnus J. Wenninger (Cambridge University Press, 1971), which contains photographs of 119 polyhedra and instructions for building them. According to H. S. M. Coxeter, the models constitute "what is almost certainly the only complete set ever made of the known uniform polyhedra."

Paperweights in the shape of cuboctahedra are quite popular: I have a crystal one made in Japan, an expensive inlaid onyx model, and a handsome little one made of stone in Mexico that cost less than a dollar. Import stores are good places to look for them.

Second day

Show Transparency 5-30. The element carbon exists in three forms. The first form, graphite, is a soft black substance that is the main ingredient of the "lead" used in pencils. The carbon atoms in graphite are arranged in layers which slide easily against each other. One layer is shown in the figure. Do you recognize the pattern of the atoms in it? (It is the 6-6-6 mosaic, the pattern used by bees in building their honeycomb!)

The second form of carbon, diamond, is the hardest substance known. In diamond, the carbon atoms are arranged in a three-dimensional network in which each atom is surrounded by four others. The lines in the figure outline the basic unit in this network. Do you recognize what it is? (The regular tetrahedron.)

Show Transparency 5-31. A third form of carbon, called the "bucky ball," was discovered in 1989. In this form, the carbon atoms appear at the corners of a semiregular polyhedron. Do you recognize which one? Add the overlay as a clue. (It is the truncated icosahedron, the "soccer ball" polyhedron.) This form of carbon was created in the laboratory but is believed to occur naturally in outer space.

In the figure, the white spheres represent carbon atoms and the black sticks connecting them represent bonds between the atoms. How many carbon atoms does a bucky ball contain? Guide your students in using the fact that the ball has 12 faces that are pentagons and 20 faces that are hexagons to find out.

Transparency 5-32 is for possible use in discussing some of the Set I exercises of Lesson 5.

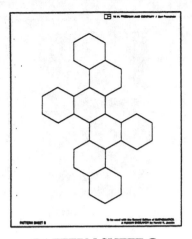

PATTERN SHEET G

Most teachers will probably want to have those students who are interested in doing the Set III exercise on filling space with truncated octahedra do it at home. If it is done in class, you may want to give each student a duplicated copy of Pattern Sheet G. Each student will need the following materials:

Five sheets of stiff paper

Pattern Sheet G (or tracing paper)

Scissors

A compass

A ruler

Tape

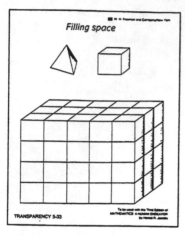

TRANSPARENCY 5-33   TRANSPARENCY 5-34   TRANSPARENCY 5-35

LESSON 6

Pyramids and Prisms

Show Transparency 5-33. The Greek philosopher Aristotle thought that, of the regular polyhedra, only two can be used to fill space: the regular tetrahedron and the cube. He was wrong about the tetrahedron. The cube is the only regular polyhedron that can fill space.

Among the semiregular polyhedra, there is also only one that can be used to fill space. Show Transparency 5-34. It is shown in this picture. Which polyhedron is it? (The truncated octahedron.) The picture shows truncated octahedra made by the St. Charles Manufacturing Company in St. Charles, Illinois. They are called "self spaces" and are for use in elementary school classrooms. The square faces are open so that a child can climb in and out of them. Inside each space is a desk and a light so that it can be used for individual study.

If some of your students have done the Set III exercise of Lesson 5, you might show some of the models they have made illustrating the special space-filling arrangement of this polyhedron.

NEW LESSON

Transparency 5-35 is for use in introducing the new lesson. This photograph shows an overhead view of a couple of very famous objects. Do you recognize what they are? The pyramid at the top, known as the Great Pyramid, was built by King Cheops in about 2600 B.C. It is situated on a plateau about five miles west of Giza. The other pyramid was built by a later Egyptian king, Chephren. These two pyramids, together with a third one located nearby, undoubtedly constitute the most well-known group of monuments in the world. On a visit to Egypt, the French general Napoleon calculated that the three pyramids together contain enough stone to build a wall 10 feet high and 1 foot thick around the entire country of France.*

Pictures and information on the construction of the pyramids is given on pages 129-139 of the book *Ancient Egypt* by Lionel Casson, which is part of Time-Life's "Great Ages of Man" series.

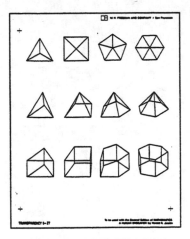

TRANSPARENCY 5-36

Transparency 5-36 with its two overlays are for use in presenting the rest of the lesson in accord with the text.

The Pyramids of Egypt, by I. E. S. Edwards (The Viking Press, 1972).

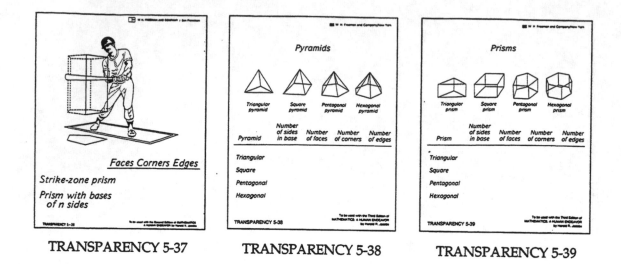

TRANSPARENCY 5-37   TRANSPARENCY 5-38   TRANSPARENCY 5-39

Review of Chapter 5

First day

Show the figure at the top of Transparency 5-37, keeping the rest of the transparency covered. What does a baseball player call the region outlined in this figure? (It is the strike zone, the region over home plate extending upward from the batter's knees to his armpits.) What mathematical name can be used to describe this shape? (It is a prism.) What shape do its bases have? (The shape of home plate, which is a nonregular pentagon.)

The following questions constitute a good review of some of the exercises in Lesson 6. Reveal the table on the lower part of the transparency and complete it as your students answer the following questions. How many faces does the strike-zone prism have altogether? (Seven.) How many faces does a prism have altogether if each of its bases has n sides? ($n + 2$.) How many corners does the strike-zone prism have? (Ten.) How many corners does a prism have if each of its bases has n sides? ($2n$.) How many edges does the strike-zone prism have? (Fifteen.) How many edges does a prism have if each of its bases has n sides? ($3n$.) What relation do the numbers of faces, corners, and edges of the strike-zone prism have? (The sum of the numbers of its faces and corners is two more than the number of edges.) Is this true of every prism? (Yes.) Why? (Because $(n + 2) + 2n = 3n + 2$ for every value of n.)

Transparencies 5-38 and 5-39 are for possible use in discussing some of the Set II exercises of Lesson 6.

The two star polyhedra pictured in the Set III exercises (page 308) make very attractive models. Detailed information is given on them on pages 90-91 and 94-95 of the *Mathematical Models* book by Cundy and Rollett. The five triangles in any one plane should be of the same color; if those in parallel planes are of the same color, only six colors are needed. In both cases, to get a rigid model, it is best to glue pyramids, one at a time, to the faces of a dodecahedron (or icosahedron) core, as illustrated in the figure in the margin of page 308 of the text.

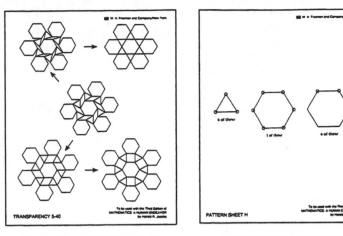

TRANSPARENCY 5-40   PATTERN SHEET H

Second day

In his delightful book *Hidden Connections, Double Meanings* (Cambridge University Press, 1988), David Wells poses the following question about the figure in the center of Transparency 5-40. Show the transparency with the other figures covered at first. Imagine that in this tessellation of hexagons and triangles, the parallelograms are merely empty spaces. Will the hinged pieces be movable? What will their extreme positions be?

The other figures on the transparency reveal the answers. The pieces are movable and, in the extreme positions, form the mosaics 3-6-3-6 and 3-4-6-4.

A more dramatic way to illustrate this than showing this transparency is to cut hexagons and triangles from heavyweight plastic of two colors and attach them with pins through tabs at their vertices. Patterns for doing this are given on Pattern Sheet H. Unfortunately, the assembly does not close perfectly on the 3-6-3-6 mosaic because pairs of pins cannot be in the same place at the same time. The arrangement is reasonably close, however, and the simultaneous movements of the pieces in going from one position to the other are fun to see.

Ask your students to write the appropriate symbols for the two mosaics formed in the extreme positions. They might also use the measures of the angles of the polygons to show that they exactly surround each point in each case. A final question concerns the intermediate positions in which parallelograms "fill" the empty spaces. Are these patterns mosaics? (They are like mosaics in that each corner point is surrounded by the same polygons in the same order. The parallelograms are not regular polygon however.)

A nice bulletin-board display when this chapter is nearly finished can be made from a set of posters designed by Joseph N. Portney. There are twelve 9-by-12-inch posters in the set showing photographs of models of an earth that is flat, cubical, dodecahedral, pyramidal, cylindrical, conical, toroidal, hyperboloidal, hyperbolic paraboloidal, oblate spheroidal, "Wegeneroidal," and "cataclysmal." Called "Earthshapes Posters," they are available from Creative Publications, 5040 West 111th Street, Oak Lawn, IL 60453.

CHAPTER

Mathematical Curves

This chapter introduces the student to curves that are ordinarily studied only in second-year algebra, trigonometry, and analysis. These curves are linked to a wide variety of natural phenomena, as well as to objects and events from daily life. Emphasis on these connections makes the study of mathematical curves exciting and concrete. It also enables students to develop a greater awareness of the importance of a knowledge of mathematics in understanding their immediate surroundings and the broader context of the universe in which we live.

The first three lessons deal with the circle, ellipse, parabola, and hyperbola; the fact that they may be viewed as conic sections is not revealed until all of them have been introduced. The representation of these curves by equations is shown in exercises in which their graphs are given. Paper-folding experiments that produce the envelope of each conic section are also included.

In the lesson on the sine curve, the student begins by constructing a sine table, which is then used to graph the curve. The remaining lessons deal with spirals and cycloids; in each case, the student begins by drawing the curve before considering some of its properties. The emphasis is on discovery and not on memorization.

TRANSPARENCY 6-1

Halley's Comet

March 30, 239 B.C.	September 9, 989
October 5, 163 B.C.	March 23, 1066
August 2, 86 B.C.	April 22, 1145
October 5, 11 B.C.	October 1, 1222
January 26, 66 A.D.	October 23, 1301
March 20, 141	November 9, 1378
May 17, 218	June 9, 1456
April 20, 295	August 25, 1531
February 16, 374	October 27, 1607
June 24, 451	September 15, 1682
September 25, 530	March 13, 1759
March 13, 607	November 16, 1835
September 28, 684	April 20, 1910
May 22, 760	February 9, 1986
February 27, 837	July 28, 2061
July 9, 912	March 27, 2134

TRANSPARENCY 6-2

TRANSPARENCY 6-3

The Circle and the Ellipse

First day

Show Transparency 6-1. This drawing shows Halley's comet as it appeared in New York City in 1910. The comet, the most famous in history, has been seen from the earth many times. Show Transparency 6-2. The dates on which it was closest to the sun, starting with an observation of it in 239 B.C., are shown on this list. Edmund Halley saw the comet in 1682 and not only correctly identified its earlier visits of 1531 and 1607, but also correctly predicted its return in 1759. Although he did not live long enough to see his prediction come true, the comet is named in his honor.

Use Transparencies 6-3 and 6-4 to introduce the rest of the lesson in accord with the text. Draw radii to the two points indicated on the circle and focal radii to the three points on the ellipse, measuring them in centimeters with a transparent ruler, as you explain the definitions of circle and ellipse. (You might tie the two ideas together by suggesting that your students imagine the center of the circle dividing into two points that move apart to become the two foci of the ellipse. Since the radius of the circle is 5 centimeters, it is obvious that the sum of the distances measured from the center of the circle to a given point on it and from the given point back to the center is a fixed number, 10 centimeters. Although the ellipse does not have a fixed radius, it is still true that the sum of the distances from one focus to a given point on the ellipse and from the given point to the other focus is also a fixed number, 10 centimeters.)

To do the ellipse-drawing experiment in the Set I exercises, each student needs the following materials:

Two sheets of graph paper, 4 units per inch or 2 units per centimeter.

String, 24-inch length*

---

*Do not use silk kite twine; not only does it tend to unravel, making the tying of knots difficult, but it is also so slippery that a pencil will continually slip out of a loop made of it.

97

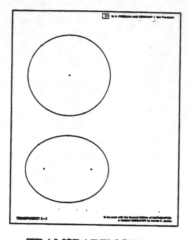

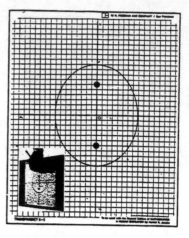

TRANSPARENCY 6-4   TRANSPARENCY 6-5

A ruler

Scissors

One 12-by-12 inch sheet of corrugated cardboard (the thicker the better)

Four thumbtacks

The thumbtacks can be distributed with
the sheet of cardboard by putting several thick-
nesses of masking tape on one of its corners
and sticking the tacks through them into the
board.

Many lengths of string can be easily cut at
one time by winding the string around the
longer dimension of a ream package of 8 $\frac{1}{2}$-
by-11-inch paper many times and then cutting
through all of the loops at once.

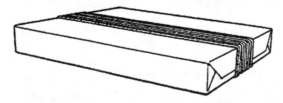

To obtain loops having the given lengths,
you might suggest the following method to
your students. Fold the string in half, putting

it next to the ruler like this (as shown in the
figure below), with the folded-over end in line
with the end of the ruler.

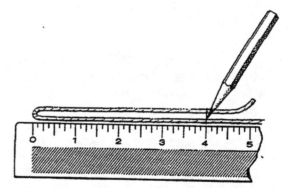

Mark the string (both upper and lower) with
your pencil at the appropriate point (4 inches
from the end for an 8-inch loop.) Take the
string away from the ruler and make a loose
knot near the two marked points. As you
tighten the knots, maneuver the string so that
the two marked points are just within it. Put
the loop back on the ruler to check to see if it is
the correct length. Finally, cut off the extra
string, being careful not to cut too close to the
knot.

Transparency 6-5 is intended for use in

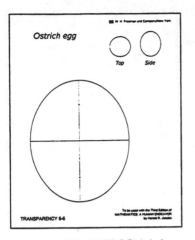

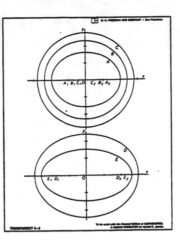

TRANSPARENCY 6-6   TRANSPARENCY 6-7

demonstrating the beginning of the ellipse-drawing experiment.

Second day

The largest eggs in the world are those laid by the African ostrich.* A typical ostrich egg weighs between three and four pounds and has an extremely thick shell that is ivory-like in texture.

Hand out duplicated copies of Transparency 6-6. The two figures at the top of this transparency show the shape of an ostrich egg as seen from the top and from the side. What curve does each shape look like? (A circle and an ellipse.)

The large figure shows the actual size of the egg. Use a ruler to measure the distance from the top to the bottom of the egg in centimeters. (15 cm.) What is this dimension of an ellipse called? (The major axis.) The surface of the egg can be generated by rotating the ellipse about this axis.

---

*If possible, this lesson should be illustrated with an actual ostrich egg. Ostrich eggs are available from stores such as The Nature Company.

What is the length of the minor axis? (12 cm.) What are the two points on the major axis called? (The foci.)

Choose a point on the ellipse and measure the distances of this point from each focus, each to the nearest tenth of a centimeter. What is the sum of the two distances? (15 cm.)

Ask your students to do this for several more points on the ellipse. The result—that each sum is the same, 15 cm—will remind them both of the definition of the ellipse and of the method used to draw it with a loop of string.

Transparency 6-7 is for use in discussing the Set I exercises of the lesson.

 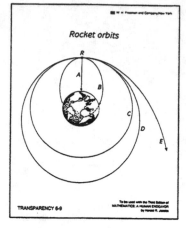

TRANSPARENCY 6-8   TRANSPARENCY 6-9   TRANSPARENCY 6-10

LESSON 2
The Parabola

First day

Show Transparency 6-8. This article, predicting a possible trip to the moon in 1965, appeared on the first page of the *Los Angeles Times* on January 23, 1955. The trip was finally made in 1969. It is interesting to note that at the time the article was written it was "still impossible to produce a single rocket that will reach orbit and escape velocities of over 18,000 m.p.h."

Show Transparency 6-9. This figure represents some of the possible orbits of a rocket fired from the earth. Suppose that the rocket has reached point R above the earth's atmosphere. If no force other than gravity is applied, the rocket might fall back to the earth, as shown by path A. If at point R the rocket is aimed parallel to the ground and given a short boost by its motor, it might travel along part of an ellipse, shown as path B, before striking the earth. Stronger boosts might put it into a circular orbit, C, or another elliptical orbit, D. If the rocket is given just enough of a boost to

enable it to escape from the earth, its path is along the curve labeled E. This curve, called a *parabola*, is the subject of the new lesson.*

Transparency 6-10 can be used to explain the definition of the parabola. Hand out duplicated copies of it and guide your students in drawing segments from the focus (located at the center of the earth) to the three points indicated on the parabola and segments from these points perpendicular to the line, measuring their lengths in centimeters. Explain that the line is called the *directrix* of the parabola.

---

*One focus of each curved path of the rocket is located at the center of the earth; in ellipse B, the other focus is above the earth. In circle C, it has moved to coincide with the center of the earth. In ellipse D, it has moved below the center of the earth. In the parabolic path, E, it has moved "infinitely far away."

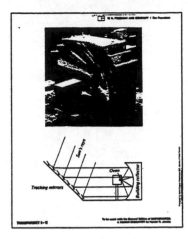

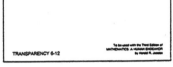

Broad jump

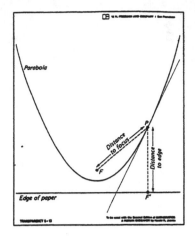

| TRANSPARENCY 6-11 | TRANSPARENCY 6-12 | TRANSPARENCY 6-13 |

Second day

A child playing with a magnifying glass outdoors on a sunny day may discover that it can be used to set a scrap of paper on fire. This suggests that sunlight might be used as a powerful source of energy in some sort of solar furnace.

Show Transparency 6-11. Such a furnace has been built in the Pyrenees mountains in France. It consists of two buildings, shown in the photograph, and an array of mirrors on a sloping hillside. The mirrors are controlled by a computer to follow the sun during the day. Sunlight is reflected from these mirrors toward the large building at the right in the photograph and diagram. The wall of the building, which appears to be "caved in," is eight stories high and is about half the area of a football field. It consists of 8,570 mirrors arranged to form a parabolic surface so that the light reflected from them is brought into focus at one point. The furnace (the smaller building in the picture) is located at this focus and has produced temperatures of more than 6,000°F.

The wonderful aspect of this solar furnace is that the energy it produces is nonpolluting and costs almost nothing. The parabola was used in its design because of its reflective property, which is explored in the Set II exercises of the lesson.

Transparency 6-12 is for possible use in discussing the broad jump exercises of Set I.

Each student needs one sheet of unlined $8\frac{1}{2}$-by-11-inch paper for the parabola paper-folding experiment (the Set III exercise). If this experiment is done in class, it may be helpful to demonstrate the procedure with a half-sheet of transparency film.

Transparency 6-13 can be used to show why the folded method works. Fold the film along the indicated line and tape it to the front of the transparency frame rather than the back so that the part of the figure below the fold can be folded upward. In showing the transparency, explain that, when a point on the edge is folded over to the focus, the fold touches the parabola at the point labeled P where the distance to the edge (the directrix) equals the distance to the focus. This agrees with our definition of a parabola: the set of all points in a plane such that the distance of each point from a fixed point is the same as its distance from a fixed line.

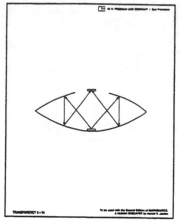

TRANSPARENCY 6-14   TRANSPARENCY 6-15

LESSON 3

The Hyperbola

REVIEW OF LESSON 2

A spectacular way to reinforce ideas about the parabola is with an illusion called Mirage. It consists of two parabolic mirrors placed face-to-face to form a covered bowl that has a circular opening in the cover.

When a small object is placed in the bottom of the bowl, a three-dimensional image appears on the cover. The image is so realistic that it is astonishing to be able to put one's hand through it.

The Mirage illusion is manufactured by Opti-Gone Associates, Box 271366, Fort Collins, CO 80527, phone (303) 282-0101, and is avail-able from the Edmund Scientific Company, 101 East Gloucester Pike, Barrington, NJ 08007, phone (609) 547-8880 and stores such as The Nature Company.

Transparency 6-14 can be used to show how the image in the Mirage illusion is formed. The object is at the focus of the upper parabola and its image appears at the focus of the lower parabola.

Transparency 6-15 is for use in discussing some of the Set II exercises of Lesson 2.

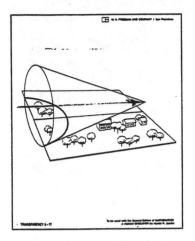

TRANSPARENCY 6-16

NEW LESSON

Transparency 6-16 can be used to introduce the lesson on the hyperbola in accord with the text. A wooden, sectioned cone is a good model for demonstrating the different conic sections. You may wish to have your students make a drawing of the sort shown here as you do the demonstration. Although it may look too difficult, I have found that, if I lead my students through it step by step, they make excellent drawings and take considerable satisfaction with their success in doing so.

Although there are a couple of methods for drawing hyperbolas that use tacks and string, they are somewhat difficult to carry out.* Producing a hyperbola by paper-folding, on the other hand, is fairly easy.

For the hyperbola paper-folding experiment, each student needs one 6-by-9-inch sheet

---

*One of them is illustrated on page 49 of *Curves and Their Properties* by Robert C. Yates (N.C.T.M., 1974) and the other on pages 207-208 of *Penrose Tiles to Trapdoor Ciphers* by Martin Gardner (W. H. Freeman and Company, 1989).

The conic sections

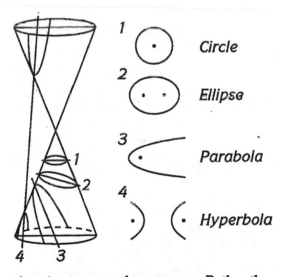

of tracing paper and a compass. Rather than handing out and collecting compasses for drawing just one circle, you may prefer to have a student assistant use a template to draw the circles on pieces of tracing paper before class.

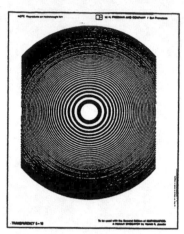

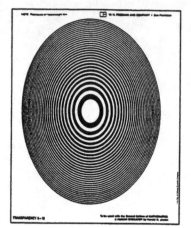

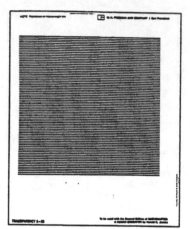

TRANSPARENCY 6-17   TRANSPARENCY 6-18   TRANSPARENCY 6-19

LESSON 4
The Sine Curve

First day

SOME "MOIRÉ CONICS"

Showing several Moiré patterns is a fascinating way to review some of the curves studied so far. Moiré patterns are produced when two periodic patterns are superimposed.

Make two copies of Transparencies 6-17 and 6-18 and one of 6-19. Show them in the following combinations.

Place 6-17 on 6-17. The pattern produced by circles placed on circles consists of parallel lines. (See Figure 1 at the right.) The distance between the lines depends on the distance between the centers of the circles and their direction depends on the relative positions of the centers.

Place 6-17 on 6-19. The pattern produced by circles placed on parallel lines consists of two sets of "ghost" circles. (See Figure 2 on the next page.) Sliding one pattern over the other causes the "ghost" circles to look like ripples in a pool.

FIGURE 1

Place 6-18 on 6-18, at first keeping the major axes of the ellipses parallel. The pattern produced by ellipses placed on ellipses in this way again consists of parallel lines. If one transparency is slowly rotated, however, hyperbolas will appear. (See Figure 3 on the next page.)

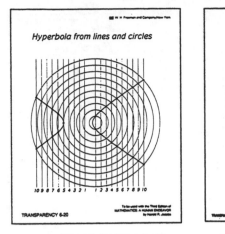

TRANSPARENCY 6-20

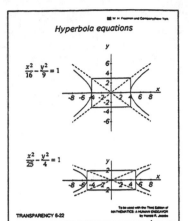

TRANSPARENCY 6-21

TRANSPARENCY 6-22

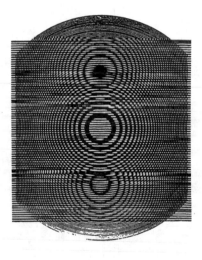

FIGURE 2

FIGURE 3

REVIEW OF LESSON 3

Transparencies 6-20, 6-21, and 6-22 are for use in discussing the exercises of Lesson 3.

NEW LESSON

To introduce the new lesson, hand out duplicated copies of Transparency 6-23. The four seasons result from the fact that the length of daylight, the time between sunrise and sunset, varies during the year.* The word *equinox* refers to the time when day and night are of equal length, that is, 12 hours each. The spring equinox occurs on about March 21 and is the

---

*A good discussion of the seasons appears on pp. 21-23 of *The Rhythms of Life*, edited by Edward S. Ayensu and Philip Whitfield (Crown Publishers, 1982).

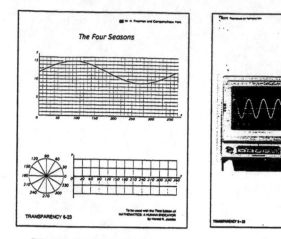

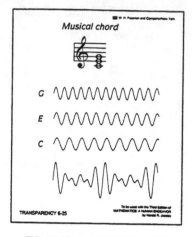

TRANSPARENCY 6-23   TRANSPARENCY 6-24   TRANSPARENCY 6-25

first day of spring. The graph at the top of the transparency has been drawn starting with this day as 0. Have your students do the following. Label the x-axis "days since the first day of spring" and the y-axis "hours of daylight"; also shade the horizontal line across the graph corresponding to 12 hours. Draw vertical lines corresponding to the first day of summer (the longest day of the year), the first day of autumn (the autumn equinox), and the first day of winter (the shortest day of the year.) Label the four seasons on the graph and explain that the curve is an example of a *sine curve*.

The other figures on Transparency 6-23 can be used to develop the sine curve in terms of the motion of a crank handle on a wheel as in the text. An especially effective way to do this is to cut a circle from heavy colored plastic, mark a point on it to represent the crank handle, and attach it with a pin to the circle on the transparency. As you turn the wheel, your students can mark the various positions of the handle with respect to the x-axis as a function of the number of degrees the wheel has turned.

Notice that if we let the radius of the wheel be 1 unit, then the y-axis of the graph extends from 1 down to -1. Label the x-axis "angle

wheel has turned," and the y-axis "sine of angle." In the first exercises of the lesson, the students draw a more precise graph of the sine curve.

Second day

If you have access to the videotape "Sines and Cosines, Part 1" from the *Project MATHEMATICS!* series and the equipment to show it, a good way to open this class is to show your students part 3 of the tape, titled "Sine Waves and Sound." Although it lasts only 3 minutes and 15 seconds, it is well worth the effort to show it. The *Project MATHEMATICS!* video series is produced by Tom M. Apostol, professor of mathematics at Caltech, and is available through both the M.A.A. and N.C.T.M.

An alternative introduction to this lesson is suggested by Transparencies 6-24 and 6-25. Explain to your students that the photograph on Transparency 6-24 is of an oscilloscope, an instrument that can be used to change sound waves into visual images. If the sound waves are very simple, such as musical notes produced by tuning forks, the image on the screen is a sine curve.

The figures on Transparency 6-25 show the oscilloscope images of the three notes of a musical chord, played both separately and together. The three sine curves provide a good way to review the ideas of amplitude and wavelength.* All three have the same amplitude, which means that they have the same loudness. The higher the frequency of the note, the shorter its wavelength. Point out also that although the curve for the chord is not a sine curve, it has a repeating pattern.

---

*Be careful not to imply that sine curves are actual pictures of sound waves. There is a correspondence between the two in amplitude and wavelength, but they are not the same. Sound waves are pressure waves in the air, whereas the waves on an oscilloscope screen are simply graphs of the varying pressure at the microphone.

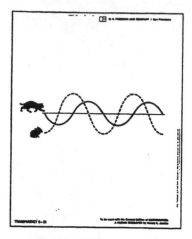

TRANSPARENCY 6-26

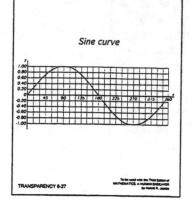

Sine curve

TRANSPARENCY 6-27

Sound wave sine curves

$y = 4\ sine\ 2x$ amplitude is
frequency is
wavelength is

$y = 2\ sine\ 5x$ amplitude is
frequency is
wavelength is

$y = 3\ sine\ \frac{1}{2}x$ amplitude is
frequency is
wavelength is

TRANSPARENCY 6-28

LESSON 5
Spirals

REVIEW OF LESSON 4

Show Transparency 6-26. Sine curves can be used to picture the interdependence of animal populations. This figure, from an article in *Scientific American*,* illustrates the changes in the numbers of rabbits and lynxes in a certain territory over a period of time. The rabbits are the chief food of the lynxes; so, when there are lots of rabbits, the lynx population increases. With more and more lynxes, the number of rabbits caught increases and so the rabbit population falls. Fewer rabbits means less food for the lynxes and so the lynx population decreases. Fewer lynxes to catch the rabbits results in an increase in the rabbit population, and so forth.

The figure shows two interlocking sine curves. How do they compare in amplitude? (The rabbit curve has a greater amplitude, which indicates that the rabbit population fluctuates more than the lynx population.)

*"Feedback" by Arnold Tustin, September 1952.

How do the curves compare in wavelength? (They are the same, because of their interdependence.)

Transparencies 6-27, 6-28, and 6-29 are for use in discussing some of the exercises of Lesson 4. It is interesting to point out in discussing exercises 6-9 of Set II that television stations broadcast the picture part of the signal by AM and the sound part by FM.

NEW LESSON

A good way to introduce the lesson on spirals is to show a chambered-nautilus shell that has been cut in half so that the chambers can be seen. Such a shell is fairly easy to find in stores and is worth looking for. Transparency 6-30 can be substituted for a shell if you do not have one.

A typical nautilus moves through about 36 chambers in the course of its life, living in each one for about a month. A good book on the chambered nautilus is *In Search of Nautilus* by Peter Douglas Ward (Simon and Schuster,

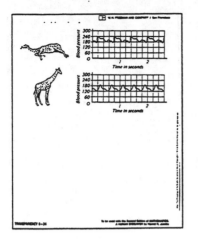

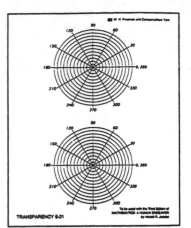

TRANSPARENCY 6-29   TRANSPARENCY 6-30   TRANSPARENCY 6-31

1988).

Rather than having your students use tracing paper for the Set I exercises of Lesson 5, you may prefer to hand out duplicated copies of Transparency 6-31. The transparency itself can be used to guide your students in getting started.

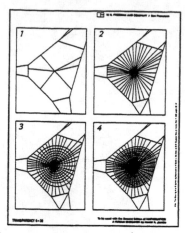 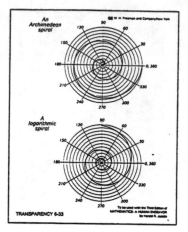 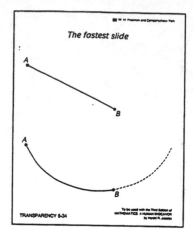

TRANSPARENCY 6-32   TRANSPARENCY 6-33   TRANSPARENCY 6-34

LESSON 6
The Cycloid

First day

REVIEW OF LESSON 5

The way in which a spider weaves its web is remarkable. These figures (on Transparency 6-32) are based on drawings made by a structural engineer as he observed a spider carrying out the various stages of construction.*

In the first stage, the spider builds the main structural lines and puts in a few radial lines to establish the center of the web.

The second stage consists of building the framework of the web by adding many more radial lines.

In the third stage, the spider starts at the center of the web and spins a spiral, which serves as a temporary scaffolding. The spider keeps the loops evenly spaced by putting one

---

*The information about the construction of the spider's web is from an article titled "The Weaving of an Engineering Masterpiece, a Spider's Orb Web" by B. E. Dugdale in *Natural History*, March 1969.

leg on the preceding loop as it makes the next one. What kind of spiral does it make? (Archimedean.)

In the final stage, the spider begins at the outside and, going in the opposite direction, spins a second spiral. This spiral has many more loops than the first one and is made of sticky thread so that insects flying into the web will be caught by it. Especially remarkable is the fact that, as the spider spins this sticky spiral, it simultaneously removes the first one, rolling it into a ball and eating it at the end. With the second spiral in place, the first one is no longer needed and so the spider, rather than wasting the material, consumes it so that it can be reused.

Transparency 6-33 is for use in checking the Set I exercises of Lesson 5.

NEW LESSON

Show Transparency 6-34, keeping the second figure covered. This figure shows a children's slide. The shortest distance from the

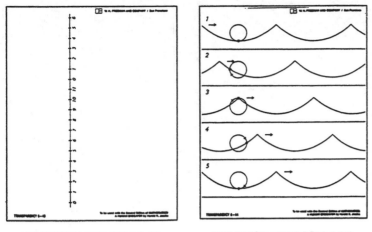

TRANSPARENCY 6-35   TRANSPARENCY 6-36

top of the slide, A, to the bottom, B, is along the slide because it has the shape of a straight line. To travel this distance takes a certain amount of time.

Now suppose that the slide were curved. What do you think would happen to the time for a child to travel from A to B? The slide would be longer; could the time be shorter?

In considering this question back in the 17th century, many mathematicians agreed that if the slide were curved, the time could be shorter. The curve turned out to be different from any that we have studied. (Reveal the second figure on Transparency 6-34.) It was discovered in 1696 by Johann Bernoulli, a Swiss mathematician and scientist, and turned out to look like this. It is part of a curve called a *cycloid.* Strangely, the cycloid is also the curve followed by a pebble stuck in a tire as it rolls along a level road.

With this introduction, your students are ready to do the cycloid experiment, for which they will need the following materials:*

One sheet of graph paper, 4 units per inch

One 3-by-5-inch file card

A small piece of masking tape

A compass

A ruler

Scissors

A paper punch (2 or 3 for the class)

Your students should have little difficulty in following the directions given in the text without help. Transparency 6-35 and the wheel on the last page of the transparency book can be used, however, if you wish to help your class in getting started. (The scale on the line is slightly compressed because the circumference of the wheel is not 12 units but $2\pi(2) \approx$ 12.57 units long.)

_____

*I have found it convenient to supply file cards on which the circles have already been drawn and to which the piece of masking tape is fixed. A

student assistant might use a circle template to prepare these cards before class. If this is done, the compasses are not needed.

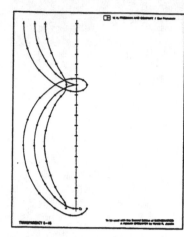
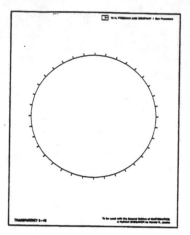
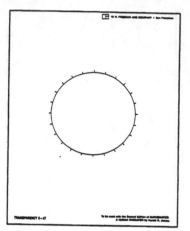

TRANSPARENCY 6-37   TRANSPARENCY 6-38   TRANSPARENCY 6-39

Second day

An ocean wave often travels a great distance from its point of origin to the beach on which it breaks. Many of the waves enjoyed by surfers on California beaches, for example, originate in the ocean east of New Zealand. Although these waves have traveled 6,000 miles, the water in them has not.

Show Transparency 6-36. The movement of the water can be observed by looking at the motion of a cork floating on its surface as a wave passes by. These drawings show a series of waves moving to the right. Notice what happens to the cork shown in the first drawing. As a wave moves past it (drawings 2 through 5), the cork travels in a circle. What kind of curve do you suppose the tops of the wave are? (Turn the transparency upside-down.) They have the form of a cycloid.*

Transparency 6-37 is for use in discussing the Set I exercises of Lesson 6.

---

*More information on the nature of ocean waves can be found in *Waves, Wind and Weather* by Nathaniel Bowditch (David McKay Company, 1977), chapter 1, and *The Book of Waves* by Drew Kampion (Arpel Books, 1989).

To save time, you may wish to hand out duplicated copies of Transparencies 6-38 and 6-39. These figures are for use in doing parts 2 and 3 of the Set II experiment.

To do the epicycloid and hypocycloid experiments, each student needs the following materials:

Two sheets of plain paper

The paper wheel made for Set I

A compass

A protractor

The paper would be the duplicated copies of Transparencies 6-38 and 6-39 if you have prepared them, in which case the compass and protractor are unnecessary.

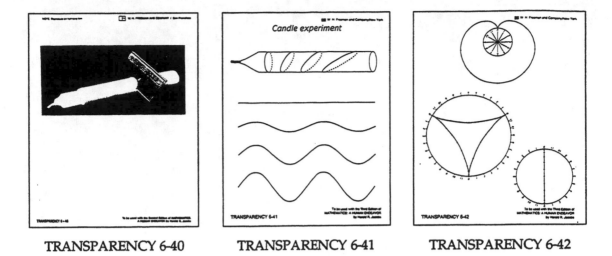

TRANSPARENCY 6-40   TRANSPARENCY 6-41   TRANSPARENCY 6-42

Review of Chapter 6

A nice way in which to close this chapter is to give a brief light show of the conic sections. Inspired by Hermann Baravalle and developed by Merwin J. Lyng, information for preparing and giving this demonstration can be found in the booklet *Dancing Curves*, published by N.C.T.M. in 1978.

Another possibility is to open with the experiment described in the Set III exercise of the review lesson. To do the candle-cutting experiment, each student needs the following materials:

1 birthday cake candle*

Scissors

Razor blade, single-edged

This experiment is a miniature version of one described and pictured on pages 233-234 of *Mathematical Snapshots* by H. Steinhaus. If your students participate, it is easiest to hand out envelopes containing the 3-by-10 centimeter strip of paper and the items listed above. Putting bright yellow mystic tape over the blunt edge of the razor blade will make it easier to see if misplaced. When the students are finished, they can put everything back into the envelope. Later, a student assistant can remove the scraps from the envelopes and prepare them to be used again. Transparency 6-40 might be used to illustrate how to wrap and cut the candle. Transparency 6-41 illustrates some possible results.

REVIEW OF LESSON 6

Transparency 6-42 is for use in discussing the Set II exercises of Lesson 6.

---

*The candle should be the plain cylindrical kind. Hallmark "Nob Hill" candles (36 per box) are available in most stationery stores if your market does not have them.

CHAPTER **7**

Methods of Counting

The seventh chapter presents a variety of counting problems whose solutions require the application of systematic methods. It also serves to help prepare the student for some of the concepts included in the following chapter on probability.

The fundamental counting principle is introduced in the first lesson, followed by two lessons dealing with the linear permutations of things that are different and of things of which some are alike. In the final lesson, the student learns how to determine numbers of combinations of a set of things. The problems that are generally presented with this kind of material are often unimaginative. "In how many orders can a dictionary, three novels, and two cookbooks be arranged on a shelf?," for example, does not seem to me to possess much potential for creating any interest among most students. An attempt has been made throughout the chapter to avoid problems of this type and to present a variety of interesting exercises that stimulate the student's natural curiosity when faced with a puzzle.

TRANSPARENCY 7-1	TRANSPARENCY 7-2

National Conference

Atlanta Falcons
Chicago Bears
Dallas Cowboys
Detroit Lions
Green Bay Packers
Los Angeles Rams
Minnesota Vikings
New Orleans Saints
New York Giants
Philadelphia Eagles
St. Louis Cardinals
San Francisco 49ers
Washington Redskins

American Conference

Baltimore Colts
Buffalo Bills
Cincinnati Bengals
Cleveland Browns
Denver Broncos
Houston Oilers
Kansas City Chiefs
Miami Dolphins
New England Patriots
New York Jets
Oakland Raiders
Pittsburgh Steelers
San Diego Chargers

License plates

A-0
RHODE ISLAND
OCEAN STATE

A-00
RHODE ISLAND
OCEAN STATE

AA-0
RHODE ISLAND
OCEAN STATE

00000
RHODE ISLAND
OCEAN STATE

TRANSPARENCY 7-1 **TRANSPARENCY 7-2** **TRANSPARENCY 7-3**

The Fundamental Counting Principle

First day

Each year the top team in the American Football Conference plays the top team in the National Football Conference in the Super Bowl. In one of the early years of the Super Bowl, a radio station in San Francisco ran a contest in which the object was to send in a postcard on which you predicted which two teams would play. You could send in as many postcards as you wanted. The prize was tickets to the game, plane fare, and cash. Soon after the contest began, however, someone pointed out that anyone who really wanted to could "beat" the contest—that is, that they could be certain of sending in a postcard on which they named the two teams correctly.

Show Transparency 7-1. At the time the contest was held, there were 13 teams in each conference. How many postcards do you suppose you would have had to send in to cover every possibility? The process of discovering that this question can be answered by multiplying 13 × 13 will prepare your stu-

dents to understand the fundamental counting principle.

The radio station couldn't afford to have everyone who sent in a card with the correct teams win the prize, so the rules were changed so that the winner was chosen by chance from all of the correct entries.

There are now 14 teams in each conference. How many postcards would you have to send in to be sure of naming the Super Bowl teams correctly if the contest were held now? (14 × 14 = 196.)

A second example of applying the fundamental counting principle is pictured on Transparency 7-2. The figure represents a set of postcards. Are there enough postcards in the picture to cover every possibility in the original Super Bowl contest? How can you find the number of cards without counting them?

A third example is suggested by the figures on Transparency 7-3. Rhode Island, the smallest state, has one of the simplest license plate numbering systems in the country. Some plates have just one letter followed by one

115

I	II	III
0 integrated	0 management	0 options
1 total	1 organizational	1 flexibility
2 systematized	2 monitored	2 capability
3 parallel	3 reciprocal	3 mobility
4 functional	4 digital	4 programming
5 responsive	5 logistical	5 concept
6 optional	6 transitional	6 time-phase
7 synchronized	7 incremental	7 projection
8 compatible	8 third-generation	8 hardware
9 balanced	9 policy	9 contingency

TRANSPARENCY 7-4

To be used with the Third Edition of
MATHEMATICS: A HUMAN ENDEAVOR
by Harold R. Jacobs

TRANSPARENCY 7-4

digit. If we assume that any letter of the alphabet can be used and any one of the 10 digits, how many of these plates are possible? (26 × 10 = 260.) Other plates used in Rhode Island are also pictured. How many of each of them are possible?

Second day

Show Transparency 7-4. Someone once invented these lists to poke fun at people who use vague and complicated words that sound impressive but don't really mean anything. The idea is to choose any three-digit number and then select the words in the three lists that correspond to the three digits.

Illustrate this by "translating" some three-digit numbers chosen by your students into three-word phrases. For example, the number 142 translates into "total digital capability." How many different three-word phrases can be made in this way? (10 × 10 × 10 = 1,000.) How many different three-word phrases could be made if each list had only 5 words instead of 10? (5 × 5 × 5 = 125.) How many could be made if each list had 20 words? (20 × 20 × 20 = 8,000.)

Near transparency 7-5 image text:

American Express Company
CREDIT CARD
EXPIRES APRIL 30, 1959
201 093 815 7
CARL W VOLCKMANN
29 NURSERY RD
HUNTINGTON STA N Y

65,000,000 MasterCards
Each has a 16-digit number

$$\frac{65,000,000}{10,000,000,000,000,000} = \frac{1}{150,000,000}$$

TRANSPARENCY 7-5

Near transparency 7-6 image text:

Slot machine

Symbol	Reel 1	Reel 1	Reel 3
Bar	3	2	1
Bell	1	5	8
Cherry	2	6	0
Lemon	0	0	4
Melon	2	2	2
Orange	5	5	4
Plum	7	3	3
7	1	1	1
(Positions)	20	20	20

TRANSPARENCY 7-6

LESSON 2

Permutations

REVIEW OF LESSON 1

Show Transparency 7-5, keeping the lower half covered. This picture shows one of the first credit cards to be issued. Issued by American Express, it was made of cardboard rather than plastic. Notice that the account number on it has 10 digits. If each digit can be any digit from 0 through 9, how many different account numbers having 10 digits are possible? (10^{10} = 10,000,000,000.) If the account numbering system used on this card were still in use, would it be possible for each person who has a credit card to have a different number? (Yes, because the population of the entire earth is less than 10 billion.)

Modern credit cards have account numbers with more than 10 digits. MasterCards, for example, have 16 digits. How many different 16-digit account numbers are possible? (10^{16}, or 10 quadrillion!) There are about 65 million MasterCards currently being used. Do you think MasterCard ever expects to have 10 quadrillion cardholders? Why do you suppose they use card numbers with so many digits? (Reveal the numbers on the lower half of the transparency.)

$$\frac{65,000,000}{10,000,000,000,000,000} \approx \frac{1}{150,000,000}$$

65 million compared to 10 quadrillion is approximately the same as 1 compared to 150 million. This means that if someone makes up a MasterCard credit card number, the chance that it matches an actual number is about 1 in 150 million. In other words, it is close to impossible to choose a 16-digit number that is an actual MasterCard number. This makes it possible to use your credit card over the phone, when the operator can't even see your card. Essentially the only way to be able to tell the operator a genuine card number is to read it from a genuine card.

Transparency 7-6 is for possible use in discussing some of the Set II exercises of Lesson 1.

117

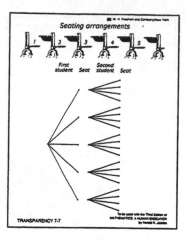

TRANSPARENCY 7-7

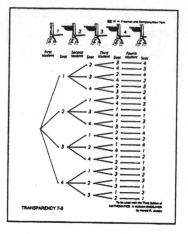

TRANSPARENCY 7-8

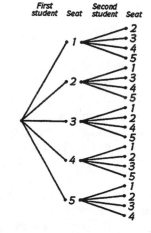

TRANSPARENCY 7-9

NEW LESSON

If the students in a class have been assigned seats, then they sit in the same arrangement every day. Suppose, instead, that each day the students sit in a *different* arrangement. If the class has ___ students and ___ seats, in how many different ways can the students be seated?*

To get an idea of how this might be figured out, consider the question for just two students and five seats. Hand out duplicated copies of Transparency 7-7. This tree diagram can be used to show all of the possibilities. Have your students fill the seat numbers into the diagram as shown at the right.

After they have noticed that there are 20 arrangements and that this number can be obtained by the fundamental counting principle (5 choices of a seat for the first student followed by 4 choices for the second student: $5 \times 4 = 20$), you can introduce the term *permutation* and the symbol $_5P_2$.

A question such as the number of possible

arrangements of four students in four seats (illustrated on Transparency 7-8), leads to the idea of *factorial*.

The number of possible seating arrangements for your own classroom situation is easy to write symbolically. If your class has n seats and r students, it is $_nP_r$. To conclude this lesson, you might prepare a transparency comparable to Transparency 7-9 to fit your class.*

---

*Use the numbers of students and seats for your own classroom situation.

*Either use a calculator to find the approximate value of $_nP_r$ or a computer program such as *Mathematica* to find its exact value.

Genetic code

```
TAGACCTCACCCTGTGGAGCCACACCCTAG
GGTTGGCCAATCTACTCCCAGGAGCAGGGA
GGGCAGGAGCCAGGGCTGGGCATAAAAGTC
AGGGCAGAGCCATCTATTGCTTACATTTGC
TTCTGACACAACTGTGTTCACTAGCAACCT
CAAACAGACACCATGGTGCACCTGACTCCT
GAGGAGAAGTCTGCCGTTACTGCCCTGTGG
GGCAAGGTGAACGTGGATGAAGTTGGTGGT
GAGGCCCTGGGCAGGTTGGTATCAAGGTTA
CAAGACAGGTTTAAGGAGACCAATAGAAAC
TGGGCATGTGGAGACAGAGAAGACTCTTGG
GTTTCTGATAGGCACTGACTCTCTCTGCCT
ATTGGTCTATTTCCCACCCTTAGGCTGCTG
GTGGTCTACCTTTGGACCCAGAGGTTCTTT
GAGTCCTTTGGGGATCTGTCCACTCCTGAT
GCTGTTATGGGCAACCCTAAGGTGAAGGCT
CATGGCAAGAAAGTGCTCGGTGCCTTTAGT
GATGGCCTGGCTCACCTGGACAACCTCAAG
GGCACCTTTGCCACACTGAGTGAGCTGCAC
TGTGACAAGCTGCACGTGGATCCTGAGAAC
TTCAGGGTGAGTCTATGGGACCCCTTGATGT
```

TRANSPARENCY 7-10

Permutations of 3 things

TAG	AC_1C_2	$G_1G_2G_3$
TGA	AC_2C_1	$G_1G_3G_2$
AGT	C_1AC_2	$G_2G_1G_3$
ATG	C_2AC_1	$G_2G_3G_1$
GAT	C_1C_2A	$G_3G_1G_2$
GTA	C_2C_1A	$G_3G_2G_1$

TRANSPARENCY 7-11

LESSON 3

More on Permutations

There are over three billion characters in the genetic code of a human being. The goal of the Human Genome Project, which started in 1990 and which hopes to finish in 2005, is to figure out the entire code.* A small sample of it is shown on Transparency 7-10. The characters show part of the code for hemoglobin, the protein that makes blood red.

As your students may recall from some of the exercises of Lesson 1, the characters of the genetic code are the letters A, C, G, and T, which stand for four kinds of small molecules. These letters form "three-letter words" which stand for different amino acids.

Consider the first three letters shown: TAG. Make a list of all of the different ways in which these three letters can be arranged. How many are there in all? (TAG, TGA, AGT, ATG, GAT, GTA; 6.) Notice that the number of ways is 3!.

Consider the next three letters shown: ACC. Can these letters also be arranged in 6 different ways? Why not?

The third line begins with GGG. These three letters can be arranged in only one way: the one shown.

Having made these observations, your students are ready to consider Transparency 7-11. (You may want to provide them with duplicated copies of it.) The duplicated letters are tagged with numbers to reveal that 3 things can be arranged in 3! = 6 ways as long as all 3 are different. If 2 things are alike, the number of arrangements is $\frac{3!}{2!} = 3$ different ways and if all 3 things are alike, the number of arrangements is $\frac{3!}{3!} = 1$ way.

Following this, have your student consider the numbers of different ways in which the following sets of letters from the hemoglobin code on Transparency 7-10 can be arranged.

1. The last 4 letters of line 1: CTAG.

---

*A good reference on this project is *The Human Blueprint* by Robert Shapiro (St. Martin's Press, 1991).

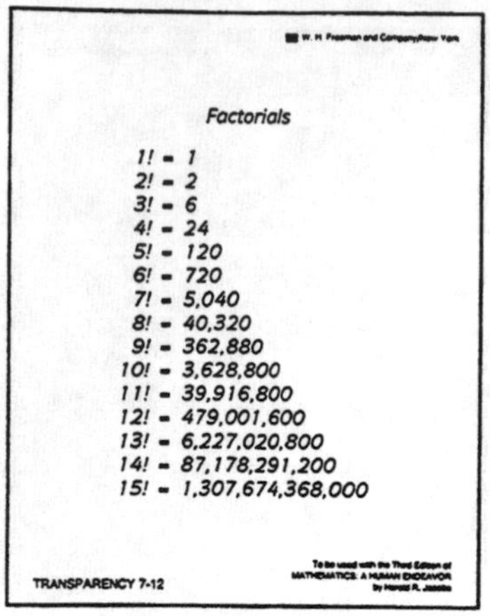

Factorials

$1! = 1$
$2! = 2$
$3! = 6$
$4! = 24$
$5! = 120$
$6! = 720$
$7! = 5,040$
$8! = 40,320$
$9! = 362,880$
$10! = 3,628,800$
$11! = 39,916,800$
$12! = 479,001,600$
$13! = 6,227,020,800$
$14! = 87,178,291,200$
$15! = 1,307,674,368,000$

To be used with the Third Edition of
MATHEMATICS: A HUMAN ENDEAVOR
by Harold R. Jacobs

TRANSPARENCY 7-12

TRANSPARENCY 7-12

2. The first 4 letters of line 1: TAGA.
3. The second 4 letters of line 1: CCTC.
4. The first 4 letters of line 2: GGTT.

The factorial table on Transparency 7-12 is for possible use in discussing some of the Set II exercises of Lesson 2.

Virginia State Lottery

1	2	3	4
5	6	7	8
9	10	11	12
13	14	15	16
17	18	19	20
21	22	23	24
25	26	27	28
29	30	31	32
33	34	35	36
37	38	39	40
41	42	43	44

TRANSPARENCY 7-13

TRANSPARENCY 7-13

LESSON 4

Combinations

Some state lotteries have had jackpots worth millions of dollars. Since there is no limit on how many tickets a person is permitted to buy, a common dream among regular lottery players is to bet on every possible combination of numbers so that winning the jackpot would be guaranteed. In 1992, a group of people apparently won the Virginia State Lottery, worth $27 million, by doing this very thing.*

Show Transparency 7-13. To win the jackpot, you had to correctly choose 6 numbers from 44. How many tickets would you have to buy to cover every possible combination?

To figure this out, let's consider a lottery in which there are only 5 numbers to choose from. Suppose you have to correctly choose 3 numbers to win the jackpot. How many tickets would you have to buy?

This is fairly easy to figure out by making a list:

    1.   1-2-3

---

*Reported in *The New York Times*, February 25, 1992.

    2.   1-2-4
    3.   1-2-5
    4.   1-3-4
    5.   1-3-5
    6.   1-4-5
    7.   2-3-4
    8.   2-3-5
    9.   2-4-5
   10.   3-4-5

If you buy 10 tickets, you will be sure to win the jackpot.

With this beginning, you can develop the rest of the lesson as in the text. The number of combinations of 3 numbers chosen from 5 is

$$_5C_3 = \frac{_5P_3}{3!} = \frac{5 \times 4 \times 3}{3 \times 2 \times 1} = 10.$$

The number of combinations of 6 numbers chosen from 44 is

$$_{44}C_6 = \frac{_{44}P_6}{6!} =$$

$$\frac{44 \times 43 \times 42 \times 41 \times 40 \times 39}{6 \times 5 \times 4 \times 3 \times 2 \times 1} = 7,059,052.$$

121

TRANSPARENCY 7-14

To be sure of winning the Virginia State Lottery when it was worth $27 million, someone would have had to buy 7,059,052 tickets; at $1 each, this would have required betting about $7 million.

Having figured this out, you might ask your students to suggest some things that would make the plan difficult to carry off or that might spoil it. (Having enough cash to buy the tickets, having enough time to buy them, having to wait for the entire prize to be paid in 20 years of installments, having to share to jackpot if more than one winning ticket is sold.)

Transparency 7-14 is for use in discussing the Set II exercises about the postal bar code in Lesson 3. Someone who looks very carefully at the bars along the edge of an envelope may notice that there are 27 bars, not 25. The two extra bars appear at each end; each is a long "frame" bar and is not part of one of the digits. More information on zip codes and the postal bar code can be found in *Reading The Numbers* by Mary Blocksma (Penguin Books, 1989).

Review of Chapter 7

First day

(Before class, use a red marking pen to fill in on Transparency 7-15 the letters shown in boldface in this diagram .)

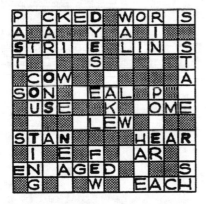

Show Transparency 7-15. Crossword puzzle contests like this one are sometimes held by newspapers. The puzzles look easy to do because most of the letters have already been filled in. They are much harder to solve than ordinary crossword puzzles, however, because the clues for most of the words are ambiguous so that it is difficult to decide what word is the most appropriate. For example, the clue for the first word across in this puzzle is: "Having _____ a small traveling bag for a trip, a woman might wish she'd chosen something bigger." The word could be either PACKED or PICKED. The rest of the clues are equally ambiguous.

The prize for correctly solving a single puzzle such as this is often several thousand dollars. Because there is usually no limit to the number of entries that a contestant may submit (as long as they are on the puzzle blanks printed in the newspaper), it might seem like a good idea to send in all of the possible solutions and hence be assured of winning the prize. How many solutions are there? To figure this out, we need to know how many squares could be filled in with more than one letter. They are the squares that are still blank. (Let your students count them. there are twenty in all.) For each clue, there are two words that

123

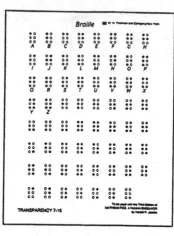

TRANSPARENCY 7-16

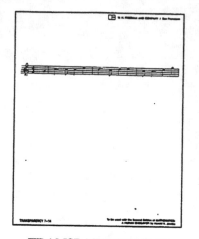

TRANSPARENCY 7-17

TRANSPARENCY 7-18

fit. This means that each of these twenty spaces can be filled in with two letters. How many ways are there to fill in all twenty spaces? ($2^{20} = 1,048,576$.) Even if you could fill in a complete puzzle blank every second, there is not enough time in an entire week to cover all of the possibilities. (A week contains 604,800 seconds.) If the newspapers cost 25¢ each, it would cost \$262,144 to get the puzzle blanks needed.

REVIEW OF LESSON 4

Transparency 7-16 may be of interest when you discuss the exercises on Braille in Lesson 4. All of the 63 characters possible are used in the Braille code; the 37 additional ones represent such things as punctuation symbols, common letter combinations, and common words.

Second day

Show Transparency 7-17. If you can read music, you can probably identify this tune. (Give your students a chance to try. If no one knows, you might give the hint that its words were written by the Scottish poet Robert Burns

and it is usually sung on New Year's Eve. It is "Auld Lang Syne.")

Suppose that, rather than seeing the music, you heard this tune but still could not identify it. A book titled *The Directory of Tunes and Musical Themes* can be used to help out in such a situation.* The user of the book does not have to know how to read music. All that is necessary is to hum the tune and decide whether each successive note goes up or down or repeats. An asterisk is written for the first note of the tune and the letters U (for up), D (for down), and R (for repeat) are used to represent the notes that follow. Demonstrate this on the transparency, writing

*URRUDDUUDRUU

below the tune. This sequence is then looked up in the book, where the name of the tune is given. Show Transparency 7-18 and locate the line for "Auld Lang Syne."

The number of tunes that can be distinguished by this simple system is remarkable.

The Directory of Tunes and Musical Themes by Denys Parsons (Spencer Brown and Company, 17 Halifax Road, Cambridge, England, 1975).

How many tunes can be coded by the first fourteen notes if each tune has a different code? (Because there is only one choice for coding the first note—an asterisk—and there are three choices for coding each of the remaining thirteen notes, the answer is $3^{13} = 1,594,323$.) This is many more than the number of tunes, 14,000, coded in the book.

CHAPTER **8**

The Mathematics of Chance

In the foreword to his fascinating book *Lady Luck: The Theory of Probability** (Doubleday Anchor Science Study Series, 1963), Warren Weaver wrote:

> The type of thinking about problems which one learns in probability theory is more than interesting, is more than fun (although I count that very valuable); it is of the highest importance. For no other type of thinking can deal with many of the problems of the modern world.

Chapter 8 introduces some basic concepts of probability theory and gives students an opportunity to apply them to a wide variety of situations. The chapter begins with a brief historical introduction, followed by a discussion of the three ways in which probabilities are determined. Several dice games are ana-

lyzed in the second lesson, which gives the student the chance to further practice the ideas introduced in the first lesson. The multiplication principle for determining the probability of several events is explained in the third lesson and applied to a variety of sets of independent and dependent events.

Two lessons on binomial probability follow, in which tree diagrams and then Pascal's triangle are used to enumerate the possible outcomes. In addition to the usual problems about the probabilities of heads and tails and boys and girls, examples of the use of probability in understanding heredity are considered.

The chapter ends with a lesson on the probabilities of complementary events and the fascinating problem of coinciding birthdays.

---

*Other especially useful sources of information on probability include *How to Take a Chance* by Darrell Huff (W. W. Norton and Company, 1959), *Probabilities in Everyday Life* by John D. McGervey (Ivy Books, 1986), *The Mathematics of Games and Gambling* by Edward W. Packel (New Mathematical Library, M.A.A., 1981), and the chapters on probability in *Elementary Statistics*, 5th edition, by Marto F. Triola (Addison-Wesley, 1992).

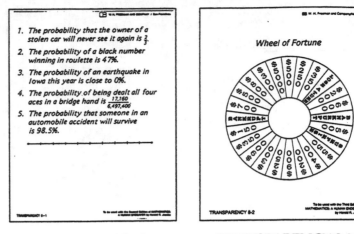

1. The probability that the owner of a stolen car will never see it again is $\frac{2}{3}$.
2. The probability of a black number winning in roulette is 47%.
3. The probability of an earthquake in Iowa this year is close to 0%.
4. The probability of being dealt all four aces in a bridge hand is $\frac{17,160}{6,497,400}$.
5. The probability that someone in an automobile accident will survive is 98.5%.

Wheel of Fortune

TRANSPARENCY 8-1   TRANSPARENCY 8-2

Probability: The Measure of Chance

Life is full of uncertainties. Here are some statements about the likelihood of various events occurring. Show the first sentence on Transparency 8-1, keeping the rest of the transparency covered, and ask your students what they think it means. Then uncover the remaining sentences one at a time, again asking about the meaning of each.

Have your students copy the scale that appears below the last statement and then direct them in numbering it in the same way that the probability scale on page 449 of the text is numbered. After recording the probabilities of the five events at the appropriate places on the scale, ask if anyone can think of an example of an event whose probability is 50%. The discussion of the meanings of the probabilities should reveal that some were determined by statistical methods (the stolen car, the automobile accident), some by mathematical computation (the roulette wheel, the bridge hand), and one by guesswork (the earthquake in Iowa).

Use the roulette-wheel and bridge-hand examples to introduce the definition of the probability of an event as the number of ways in which the event can occur divided by the total number of ways. (In the bridge-hand example, there are 17,160 possible bridge hands that contain all four aces and 6,497,400 possible bridge hands altogether.)

Transparency 8-2 can be used for additional examples of how to determine probabilities. It shows the money wheel used on the television show *Wheel of Fortune*. All of the sections of the wheel are the same size so that each one is equally likely to turn up as the wheel is spun. How many sections does the wheel have? (24.) What is the probability of the wheel ending up on each of the following?

$5,000. $\left(\frac{1}{24}.\right)$ $500. $\left(\frac{3}{24} = \frac{1}{8}\right)$

Bankrupt. $\left(\frac{2}{24} = \frac{1}{12}\right)$

Money. $\left(\frac{20}{24} = \frac{5}{6}\right)$

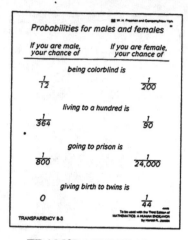

TRANSPARENCY 8-3

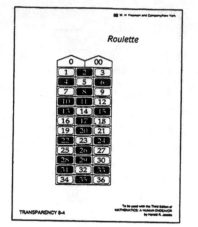

TRANSPARENCY 8-4

TRANSPARENCY 8-5

LESSON 2

Dice Games and Probability

REVIEW OF LESSON 1

Show Transparency 8-3. These probabilities are from a book by Peter Mason titled *Half Your Luck!* (Penguin Books, 1986.) Who is more likely to be colorblind: a male or a female? What does the probability $\frac{1}{12}$ mean? (That one male out of twelve is colorblind.) Who is more likely to live to a hundred? Which event is impossible? Of the possible events, which is least probable? What does the probability $\frac{1}{24,000}$ mean?

Transparency 8-4 shows the layout on which players place their chips to make bets in the game of roulette and is for possible use in discussing the exercises in Lesson 1 on roulette.

NEW LESSON

To introduce the new lesson, you might show Transparency 8-5. The dice in these

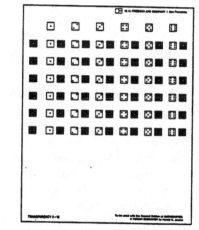

TRANSPARENCY 8-6

photographs are very old. The one at the top was found in an Egyptian tomb and is more than four thousand years old. The two at the bottom are from Italy and are dated at about the seventh century B.C. The markings on all three are nearly identical with those on modern dice. Transparency 8-6 can be used to develop the rest of the lesson in accord with the text.

128   Lesson Plans

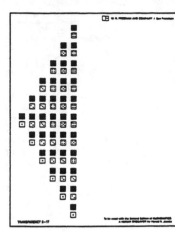

TRANSPARENCY 8-7

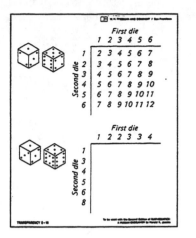

TRANSPARENCY 8-8

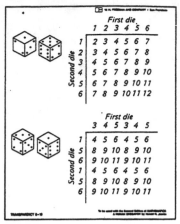

TRANSPARENCY 8-9

LESSON 3

Probabilities of Successive Events

First day

REVIEW OF LESSON 2

Show Transparency 8-7 (sideways). You know that when two dice are thrown, they can turn up in 36 different ways. The ways are shown in this figure, arranged according to the sums they produce. (Write the sums 2 through 12 in a row under the columns and then the numbers of ways of producing them in a second row.)

You also know that the dice can be numbered in other ways to produce different results. Show Transparency 8-8. Suppose that, instead of each die being numbered 1, 2, 3, 4, 5, 6, a special pair of dice is made with one die numbered 1, 2, 2, 3, 3, 4 and the other die 1, 3,

---

*These dice were invented by Col. George Sicherman of Buffalo, New York, and are discussed by Martin Gardner in Chapter 19 of his book *Penrose Tiles to Trapdoor Ciphers* (W. H. Freeman and Company, 1989).

4, 5, 6, 8.* No one would mistake these dice for normal ones for very long because of the face numbered 8.

The first table shows the sums that turn up with normal dice. Copy and complete the second table to show the sums that turn up with the special dice. What is interesting about the results? (These dice give the same sums, with the same frequencies, as do normal dice.) It has been proved that the numbering on these dice is the only one other than the normal numbering that will result in the normal probabilities.

Transparencies 8-9, 8-10, and 8-11 are for use in discussing some of the exercises of Lesson 2

NEW LESSON

The most popular card game in the United States is poker. It was first played in the nineteenth century on riverboats on the Mississippi. The people who created poker ranked the different possible hands in value accord-

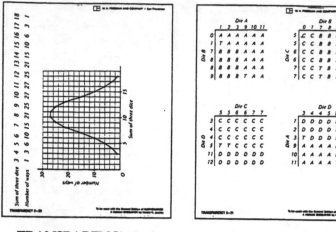

TRANSPARENCY 8-10   TRANSPARENCY 8-11   TRANSPARENCY 8-12

ing to the probabilities of being dealt them. Show Transparency 8-12.

The top ranking hand, a royal flush, consists of an ace, king, queen, jack and ten, all of the same suit. Show Transparency 8-13. One way to find the probability of getting a royal flush is to compare the number of ways of getting a royal flush, 4 (circle them on the transparency) to the total number of ways of being dealt a hand: $_{52}C_5 = 2,598,960$.

$$\frac{4}{2,598,960} = \frac{1}{649,740}$$

Another way to find this probability is based on finding the probability of being dealt each card. The first card can be any one of 20 (the 20 cards in the 4 hands you circled on the transparency) out of the 52: $\frac{20}{52}$. The second card can be any one of the remaining 4 top cards of the same suit as the first one: $\frac{4}{51}$. The third card must be one of the remaining 3 top cards of the same suit: $\frac{3}{50}$. Reasoning in the same way, the probabilities of being dealt the fourth and fifth cards of a royal flush are $\frac{2}{49}$

and $\frac{1}{48}$.

If these probabilities are multiplied,

$$\frac{20}{52} \times \frac{4}{51} \times \frac{3}{50} \times \frac{2}{49} \times \frac{1}{48} =$$

$$\frac{480}{311,875,200} = \frac{1}{649,740},$$

the result is the same as before. This suggests that probabilities can be multiplied in the same way as numbers of choices are multiplied.

In the second method, the probability of being dealt each successive card in a royal flush is affected by being dealt the previous one. Events that are influenced by other events are said to be dependent.

To illustrate the idea of independent events, consider the following. Which do you suppose is more likely: that you are dealt a royal flush in a single deal of the cards, or that, in five successive deals, you are dealt an ace as your first card each time?

Since there are 4 aces in the deck of 52 cards, the probability of being dealt an ace is $\frac{4}{52}$. Each time the deck is redealt, the probability is

TRANSPARENCY 8-13   TRANSPARENCY 8-14   TRANSPARENCY 8-15

the same: $\frac{4}{52} = \frac{1}{13}$. The probability of getting an ace as the first card of five successive deals is

$$\frac{1}{13} \times \frac{1}{13} \times \frac{1}{13} \times \frac{1}{13} \times \frac{1}{13} = \frac{1}{371,293}.$$

In this case, the probability of being dealt each successive ace is *not* affected by being dealt the previous one. Events that have no influence on each other are said to be *independent*.

Comparing the two probabilities,

$$\frac{1}{649,740} \text{ and } \frac{1}{371,293},$$

reveals that being dealt aces as your first cards in five successive deals is more likely than being dealt a royal flush.

Second day

Show Transparency 8-14. Here is a true story of how the theory of probability came to the rescue of a person who was nearly swindled out her winnings at the race track.

Rose Grant, a Brooklyn nurse, went to Belmont Park early one day to make two-dollar "triple bets" on the day's nine races. A "triple bet" is one in which the bettor must pick the first-, second-, and third-place horses in exact order.

Grant had to leave the track before the races began; so she wrote down her three choices for each race on a piece of paper and gave it with a twenty-dollar bill to one of the employees of the track who promised to place the bets for her. Show Transparency 8-15.

The next day she learned that her bet on the ninth race had won and was worth $5,050. When she went to the track to pick up her winning ticket, the employee who was supposed to have placed the bets said that she had not had a chance to do so and handed her back the paper with the choices and a twenty-dollar bill. Grant suspected that she had been cheated and filed a complaint with the track.

A computer printout of bets placed that day revealed that someone had made bets on the nine races identical with those on Grant's list at Window 18. When a man went to the track several days later to try to cash a winning ticket on the ninth race at that window, he was arrested. When the matter was brought to court, the probability of two strangers deciding on their own to bet the same 27 choices at the same window on the same day was shown to be so small that the jury found the track employee and the man who tried to cash the winning ticket guilty of grand larceny.

Suppose that there were ten horses entered in the ninth race. If someone bet on the horses at random, what is the probability that they would name the winning horse correctly? $\left(\frac{1}{10}\right)$ Having named the winning horse, what is the probability of naming the second-place horse correctly? $\left(\frac{1}{9}\right)$ The third-place horse? $\left(\frac{1}{8}\right)$ All three horses? $\left(\frac{1}{10} \times \frac{1}{9} \times \frac{1}{8} = \frac{1}{720}\right)$ With many bets being placed at the track, this probability indicates that it is not surprising that more than one person might choose all three horses correctly.

The damning evidence came with the other eight races. If there were ten horses in each race, how would you figure out the probability that two bettors would make the same triple bets in all nine races?

$$\left(\left(\frac{1}{720}\right)^9 \approx \frac{1}{52,000,000,000,000,000,000,000,000}\right)$$

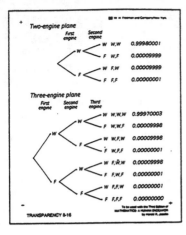

TRANSPARENCY 8-16

Binomial Probability

Several years ago, all four engines of a jet plane approaching Los Angeles International Airport failed at the same time. The plane fell 6,000 feet and then a "miracle" happened. The pilot was able to restart the engines and the plane landed safely.

For many years the Federal Aviation Administration has required that airplanes making transatlantic flights have at least three engines. Some planes, such as the Boeing 767, have just two engines. These planes have the advantage of requiring less fuel, but are not as safe. To find out how much less safe, the F.A.A. used the mathematics of probability.*

Show Transparency 8-16 with a clear sheet of film taped over it. (You may want to hand out duplicated copies of the transparency so that your students can fill in the probabilities as well.) These tree diagrams show the possible ways in which the engines of two- and

---

*This exercise is based on an example by Mario Triola in his book *Elementary Statistics*, fifth edition (Addison-Wesley, 1992).

three-engine planes can work or fail. Suppose the probability that an engine works is 90%. If this is true, what is the probability that both engines of a twin-engine plane work? ($0.9 \times 0.9 = 0.81$, or 81%.) If the probability that an engine works is 90%, what is the probability that it *fails*? (10%.) What is the probability that both engines of a twin-engine plane fail? ($0.1 \times 0.1 = 0.01$, or 1%.)

Now consider the two cases in which one engine works and the other fails, adding these numbers to the diagram as shown below:

W, W	$0.9 \times 0.9 = 0.81$, or 81%
W, F	$0.9 \times 0.1 = 0.09$, or 9%
F, W	$0.1 \times 0.9 = 0.09$, or 9%
F, F	$0.1 \times 0.1 = 0.01$, or 1%

What do the probabilities of all four cases add up to? (1, or 100%.) What is the probability that exactly one engine will fail? (0.18, or 18%.)

Now consider the probabilities for the eight cases for a three-engine plane, as shown at the top of the next page:

W, W, W	$0.9 \times 0.9 \times 0.9 = 0.729$, or 72.9%
W, W, F	$0.9 \times 0.9 \times 0.1 = 0.081$, or 8.1%
W, F, W	$0.9 \times 0.1 \times 0.9 = 0.081$, or 8.1%
W, F, F	$0.9 \times 0.1 \times 0.1 = 0.009$, or 0.9%
F, W, W	$0.1 \times 0.9 \times 0.9 = 0.081$, or 8.1%
F, W, F	$0.1 \times 0.9 \times 0.1 = 0.009$, or 0.9%
F, F, W	$0.1 \times 0.1 \times 0.9 = 0.009$, or 0.9%
F, F, F	$0.1 \times 0.1 \times 0.1 = 0.001$, or 0.1%

What is the probability that exactly one engine of a three-engine plane will fail? ($3 \times 8.1\% = 24.3\%$.) The probability that two of the three engines will fail? ($3 \times 0.9\% = 2.7\%$.)

Having done all this, explain that the probabilities in a situation in which there are two possible outcomes for each part (in this case, that each engine works or fails), are called *binomial*.

Finally, explain that the probability we have assumed for a jet engine working, 90%, is actually much too low. The true probability is closer to 99.99%. Remove from Transparency 8-16 the sheet of film on which you have made the previous calculations and add the overlay to reveal the more accurate calculations. (The probabilities for the three-engine plane have been rounded to eight decimal places to make them easier to compare to the probabilities for the two-engine plane.) If we assume that a plane will crash only if all of its engines fail, then, with these probabilities, the difference in likelihood of a two- and three-engine plane crashing seems very small. The F.A.A. decided to change its regulations to allow twin-engine planes to be used on transatlantic flights on the basis of this type of reasoning.

TRANSPARENCY 8-17

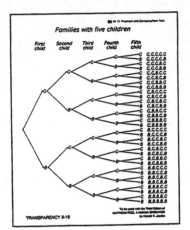

TRANSPARENCY 8-18

TRANSPARENCY 8-19

Pascal's Triangle

Show Transparency 8-17. This remarkable family, the Thomas V. Brennans of Oak Park, Illinois, consists of five consecutive girls followed by six boys. (Now cover the part of the picture showing the boys.) At one point, the family had five children, all daughters. What is the probability of this occurring?

Hand out duplicated copies of Transparency 8-18. This tree diagram shows all of the possibilities of girls and boys in a family with five children. Refer to it to copy and complete the following table.

Number of girls	0	1	2	3	4	5
Number of possibilities	(1)	(5)	(10)	(10)	(5)	(1)

What is the probability that in a family with five children, all five are girls? $\left(\frac{1}{32}\right)$ A tree diagram for the possibilities of girls and boys in a family with eleven children would be hard to draw. In this lesson, we will consider an easier way to find them.

The numbers of possibilities in the second line of our table for five children appear in a remarkable pattern of numbers called "Pascal's Triangle." Guide your students in constructing the first 11 rows of the triangle. Include the number of each row at its left and the sum of the numbers in the row at its right.

After your students have constructed Pascal's Triangle, point out that the fifth row of the triangle contains the same numbers that we counted from the tree diagram for five cildren. The first row shows the possibilities of girls and boys in a family with one child: either 1 girl or 1 boy. The second row shows the possibilities in a family with two children, the third row, a family with three children, and so forth. Transparency 8-19 may be useful in making this clearer.

The eleventh row of the triangle shows the possibilities in a family with eleven children. How many possibilities are there altogether? (2,048.) How many of these have exactly five girls and six boys? (462.) What is the probability that a family with eleven children has

135

exactly five girls and six boys? $\left(\dfrac{462}{2,048} \approx 23\%.\right)$

Is this event surprising? (No.) What is it about the family in the photograph (on Transparency 8-17) that *is* surprising? (The order in which the girls and boys were born.) What is the probability of this *particular* family occurring? $\left(\dfrac{1}{2,048} \approx 0.05\%.\right)$

In discussing the exercises of Lesson 4, your students should notice the appearance of the appropriate rows of Pascal's Triangle in the various tables they have constructed.

The Birthday Problem: Complementary Events

If you have at least twenty-five students in your class, an appropriate way to begin is with the "Coinciding Birthdays" problem. How likely do you suppose it is that two of you have the same birthday? Would you bet in favor of or against it? Because there are 365 days in a year (ignoring leap year), the usual assumption is that, if there are, say, thirty-two people in the room, the probability of two people sharing the same birthday is less than 10%.

Ask each student to tell his or her birthday as the others listen to hear if their birthdays are called. After doing this, Transparency 8-6 can be used to introduce the idea of probabilities of complementary events by considering the probabilities of getting a seven and not getting a seven when two dice are thrown.

Transparency 8-20 is for use in discussing the "Coinciding Birthdays" problem. There is space in the table for you to add the percents listed at the right.

Number of people	Probability that all birthdays are different	Probability that there are two people who share the same birthday
2	$\approx 100\%$	0%
3	$\approx 99\%$	1%
4	$\approx 98\%$	2%
5	$\approx 97\%$	3%
6	$\approx 96\%$	4%
7	$\approx 95\%$	5%
8	$\approx 93\%$	7%
9	$\approx 91\%$	9%
10	$\approx 88\%$	12%

If you have a class that is too small for the birthdays problem to be feasible, your students will enjoy an activity by Richard S. Kleber presented in his article "A Classroom Illustration of a Nonintuitive Probability" in the May 1969 issue of *The Mathematics Teacher*. The idea is to ask each student to choose a number at

137

random from a set of numbers that you specify. By selecting an appropriate set, you can make it quite probable that two of your students will choose the same number. The adjoining table lists the approximate number of numbers from which to choose so that the probability of the same number being chosen twice is either 75% or 90%.

For example, if eighteen students are present in your class, you can ask each student to choose any whole number between 1 and 115. Then have each student tell his or her number to see if there are any duplications. The probability that two students will have chosen the same number is 75%. You might then repeat the exercise, this time narrowing the choice to any whole number between 1 and 70 and telling the students that "they can't help themselves"; it is extremely likely (a 90% probability) that two will choose the same number.

My students are always fascinated, as well as exasperated, by this "game."

Number of people	75% probability	90% probability
10	35	20
11	45	25
12	50	30
13	60	40
14	70	45
15	80	50
16	90	55
17	105	65
18	115	70
19	130	80
20	145	90
21	160	100
22	175	105
23	190	115
24	205	125
25	225	140
26	245	150
27	260	160
28	280	175
29	300	185
30	325	200
31	345	210
32	370	225
33	390	240
34	415	255
35	440	270
36	465	285
37	490	300
38	520	320
39	545	335
40	575	350

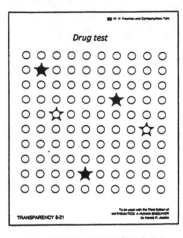

TRANSPARENCY 8-21

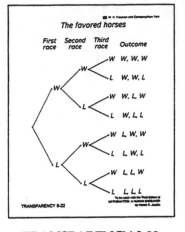

TRANSPARENCY 8-22

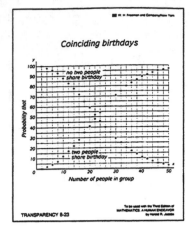

TRANSPARENCY 8-23

Review of Chapter 8

First day

More than half of all companies in the United States now test people who apply for jobs with them for drug use. For some of the people being tested, the results are not correct. That is, a test may indicate that someone uses drugs when they actually do not. Such a result is called a "false positive." A test may also indicate that someone does not use drugs when they actually do: this is called a "false negative."*

Show Transparency 8-21. Suppose that one hundred job applicants take a drug test and that the results are as shown. If the white circles represent the people who do not take drugs and the gray circles represent those who do, how many of the one hundred people take drugs? (3.) The stars identify the people for whom the test seems to indicate the use of

_____

*This exercise is based on an example by Mario Triola in his book *Elementary Statistics*, fifth edition (Addison-Wesley, 1992).

drugs. How many "false positives" are shown? (2.)

According to the results shown in the figure, what is the probability that, if a person takes drugs, the test will reveal it? ($\frac{3}{3}$ = 100%.)

What is the probability that if a person does *not* take drugs, the test will indicate that they do not? ($\frac{95}{97} \approx 98\%$.)

These probabilities make the test seem to be extremely reliable. There is a problem with the test, however. Of the people for whom the test indicates that they use drugs, what is the probability that they do not? ($\frac{2}{5}$ = 40%.)

Transparencies 8-22 and 8-23 are for use in discussing some of the exercises of Lesson 6.

Air Florida contest

Number of cards sent in	Probability of not winning a prize	Probability of winning a prize
1	96%	4%
2	91%	9%
3	87%	13%
4	83%	17%
5	79%	21%
6	76%	24%
7	72%	28%
8	69%	31%
9	66%	34%
10	63%	37%
11	60%	40%
12	58%	42%
13	55%	45%
14	53%	47%
15	50%	50%
16	48%	52%
17	46%	54%
18	44%	56%
19	42%	58%
20	40%	60%
21	38%	62%
22	36%	64%
23	35%	65%
24	33%	67%
25	32%	68%

TRANSPARENCY 8-24

To be used with the Third Edition of
MATHEMATICS: A HUMAN ENDEAVOR
by Harold R. Jacobs

TRANSPARENCY 8-24

Second day

Every year companies run sweepstakes and contests with prizes totaling many millions of dollars. Some of the people who enter these contests apply mathematics in trying to win them.

Jeff Sklar, a computer programmer, claims that he has developed a foolproof method based on probability for comparing the chances of winning a contest to the cost of entering it. He used his method on a contest run by Air Florida in which the airline distributed game cards with round-trip flights to Florida as prizes.*

There were 267,000 cards circulated and 12,000 prizes. Given that each card had an equal chance of winning a prize, what was the probability that if you sent in one card, you would win a prize? ($\frac{12,000}{267,000} = \frac{12}{267} \approx 0.045$, or 4.5%.) What was the probability that if you

---

*Reported in the article "Ways to Win Those Contest Jackpots" by Malcolm N. Carter in *Money* magazine, May 1983.

sent in one card, you would not win a prize? (0.955, or 95.5%.)

Suppose you sent in two cards. What is the probability that neither card would win a prize? ($0.955 \times 0.995 \approx 0.912$, or 91.2%.) What is the probability that if you sent in two cards, you would win at least one prize? ($1 - 0.912 = 0.088$, or 8.8%.)

Show Transparency 8-24. This situation is similar to the duplicate birthday problem. As the number of people in a group increases, the probability that at least two of them share the same birthday increases. As the number of cards a person sends in to the contest increases, the probability that at least one of the cards wins a prize increases.

According to the table, what is the probability of winning a prize if you send in 10 cards? (37%.) How many cards would you have to send in to have about even chances of winning a prize? (15.)

Jeff Sklar, the fellow who figured all of this out for the Air Florida contest, decided to send in 600 cards. As a result, he won 23 prizes. Since he did not want 23 round trips to Florida, he sold them for approximately $10,000.

An Introduction to Statistics

In their preface to the 1981 Yearbook of the National Council of Teachers of Mathematics, *Teaching Probability and Statistics*, Albert P. Shulte and James R. Smart listed several reasons for the inclusion of statistics and probability in the school curriculum.

> They provide meaningful application of mathematics. . . , they give us some understanding of the statistical arguments, good and bad, with which we are continually bombarded, they help consumers distinguish sound use of statistical procedures from unsound or deceptive uses, and they are inherently interesting, exciting, and motivating topics for most students.

The goals of the ninth chapter are to acquaint the student with some of the basic tools of statistics and to show how easily these tools can be used to give a misleading picture of what they are supposed to represent. We are constantly influenced by statistics in making consumer choices and economic and political decisions, and, from this standpoint, this chapter is perhaps the most "practical" unit of the course.

In the first lesson, the students learn how to organize data by making frequency distributions and how to draw and interpret histo-grams. The techniques introduced in this lesson are applied in the second lesson to the breaking of ciphers and codes.

The third lesson deals with the three common measures of central tendency: the mean, the median, and the mode. In the exercises in this lesson, the students are led to discover some of the advantages and disadvantages of each measure in representing various sets of data. Two measures of variability, the range and standard deviation, are introduced in the fourth lesson, together with the normal curve and its relation to binomial probabilities.

The fifth lesson, on statistical graphs, not only gives the student an opportunity to construct several basic types, but also emphasizes the ease with which even very simple graphs can be misinterpreted. The final lesson presents a brief look at the technique of sampling, the most important component of the statistical process, and, as in the preceding lessons, includes a variety of real examples. The importance of random sampling is stressed in this lesson with many examples of how errors and bias can occur when sampling is not random. Because these are subtle and important ideas, some extra care in classroom preparation is helpful on this material.

LESSON 1

Organizing Data: Frequency Distributions

First day

An interesting way to introduce the first lesson is with a set of weather data, perhaps annual rainfall or annual snowfall, for your local area. These data can be obtained from libraries and weather buréaus. The following dialogue illustrates what you might do.

Anyone who has lived in [name of your city or county] for some time knows that the amount of rain varies considerably from year to year. (Show a transparency of appropriate weather data for the past 50 years in your area.)

What is the largest amount of rainfall for a year shown in this list? What was the smallest amount in a year? There are so many numbers in the list that it is hard to "see" them all at once. One way in which they can be organized is with a frequency distribution. Guide your students in setting up a table of equal intervals and making a tally of the amounts of rainfall as you read the data.

After they have done this, have them add a column of frequencies to the table and check to see that the sum of the frequencies is 50. Finally, have your students make a histogram from the frequency distribution. When they have done this, they are ready to read the lesson and begin the exercises. (If you assign all of the exercises, it is probably best to spend two days on this lesson.)

ALTERNATE CLASS OPENER

Tell your students to take out a sheet of paper in preparation to take a short true-or-false test. Have them number from 1 to 10 and tell them that the test is very difficult and, since they have not studied for it, they will have to guess the answers. In fact, since they are guessing the answers, there is no need for you to bother to read the questions!

After your students have written down their answers, have them mark them right or wrong as you read the following answers.

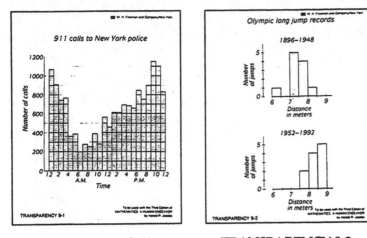

TRANSPARENCY 9-1   TRANSPARENCY 9-2

1. True.   5. False.   9. True.
2. False.   6. True.   10. False.
3. True.   7. True.
4. False.   8. False.

Explain that in this experiment, we have simulated an actual test and that we will now look at the results as a teacher might.

A common way of compiling the students' scores is to organize them in a frequency distribution. Have your students set up a table of scores ranging from 10 down to 0 and then guide them in making a tally of the scores as each student tells how many of the "questions" he or she answered correctly. Complete the table with a column of frequencies and finally make a histogram of the scores.

Second day

Show Transparency 9-1. This histogram shows the frequency distribution of 911 telephone calls to the New York City police department during a Sunday in August.* At what times were there more than 700 calls per hour? (Be-

*Adapted from "Improving the effectiveness of New York City's 911" by R. C. Larson in *Analysis of*

tween midnight and 4 A.M. and between 6 P.M. and midnight.) At what times were there less than 400 calls per hour? (Between 4 A.M. and 11 A.M.)

The police force is divided into three shifts: the first from midnight to 8 A.M., the second from 8 A.M. to 4 P.M., and the third from 4 P.M. to midnight. If this histogram is typical of the emergency calls received during a given day, how should the number of police assigned to each shift compare? (The heights of the bars suggest that more police are needed during the third shift than either of the other two and that the second shift needs the least number of police.)

Statistical information of this sort is often helpful in allocating human resources. Organizing such information in the form of frequency distributions or histograms makes it easier to interpret.

Transparency 9-2 is for use in discussing exercises 6-10 of Set I.

Public Systems, edited by A. W. Drake, R. L. Keeney, and P. M. Morse (M.I.T. Press, 1972) and *Beginning Statistics with Data Analysis* by Frederick Mosteller, Stephen E. Fienberg, and Robert E. K. Rourke (Addison-Wesley, 1983).

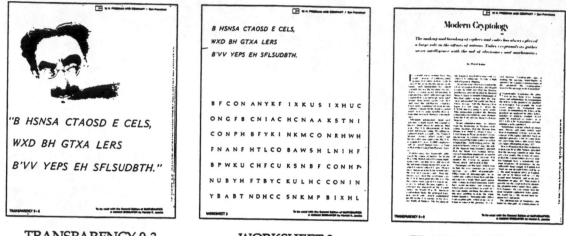

LESSON 2

The Breaking of Ciphers and Codes: An Application of Statistics

Show Transparency 9-3. This is an example of a cryptogram, a statement written in code. It is something that Groucho Marx once said and is so brief that you might think that it could not be solved without knowing the code. However, there are patterns in language, like patterns in mathematics, that make it possible to figure out what the message says.

Hand out duplicated copies of Worksheet 3. (The code on the lower half of the sheet is part of the new assignment.) The punctuation and spacing between words give some clues. Can you figure out what any of the letters might stand for? (B'VV must be I'LL, so B is I and V is L; the fourth word, then, must be A. Someone may say that BH and EH are IS and AS, but they could also be IN and AN or IT and AT.) A letter that is used very frequently in English is E. Which letter do you suppose might be E? (S, both because there are quite a few S's and because it is the final letter of several words.) With a little direction from

you and some time, your class should be able, working together, to discover that the cryptogram says: I NEVER FORGET A FACE, BUT IN YOUR CASE I'LL MAKE AN EXCEPTION.

Edgar Allen Poe, who was interested in the writing and breaking of codes, once said that he thought it very unlikely that a code could be invented that a clever person could not solve.

Show Transparency 9-4. The science of codes and codebreaking is called cryptology. According to an article on cryptology in *Scientific American* (July 1966), "the making and breaking of ciphers and codes has always played a large role in the affairs of nations." It is also playing an increasingly important role in the protection of information sent by electronic mail. According to Martin Gardner, "Government agencies and large business will presumably be the first to make extensive use of electronic mail, followed by small businesses and private individuals. When this

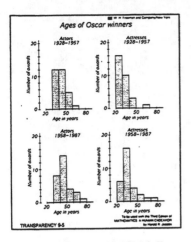

TRANSPARENCY 9-5

starts to happen, it will become increasingly desirable to have fast, efficient ciphers to safeguard information from electronic eavesdroppers. A similar problem is involved in protecting private information sorted in computer memory banks from snoopers who have access to the memory through data-processing networks."*

In his definitive book *The Codebreakers* (Macmillan, 1967), David Kahn wrote, "The cryptology of today is saturated with mathematical operations, mathematical methods, mathematical thinking. In practice, it has become virtually a branch of applied mathematics." The new lesson demonstrates one example of this: how the frequency distribution can be applied to the breaking of simple codes.

The students should now be ready to read Lesson 2 and begin the exercises. If you have given them copies of Worksheet 3, they can solve the cipher in Set II on it, rather than having to copy it on their own paper. (The

cipher is also reproduced on Transparency 9-8.)

Transparency 9-5 is for use in discussing exercises 1-7 of Lesson 1, Set II.

---

*From *Penrose Tiles to Trapdoor Ciphers* by Martin Gardner (W. H. Freeman and Company, 1989). Students who would like to learn more about recent developments in cryptology will be interested in Chapters 13 and 14 of this book.

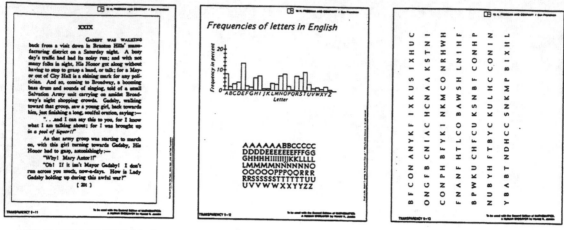

TRANSPARENCY 9-6   TRANSPARENCY 9-7   TRANSPARENCY 9-8

LESSON 3
Measures of Central Tendency

First day

REVIEW OF LESSON 2

Show Transparency 9-6. This transparency shows a page from the novel *Gadsby* by Ernest Wright. Published in Los Angeles in 1939, the book is remarkable because it violates the statistics of the English language. Can you figure out why? (The most frequently used letter in English, E, does not appear in it at all.) This is especially remarkable when you realize that *Gadsby* contains more than 50,000 words without even one appearance of such common words as "he," "she," "they," "be," "are," "were," "there,", "then," "when," or "the." More information on *Gadsby* is given by David Kahn in *The Codebreakers* (pages 739-740.)

Transparencies 9-7 and 9-8 are for use in discussing the exercises of Lesson 2. Call on a student who has deciphered the message in Set II to read it to the class. You should point out that this kind of cipher is far too simple to

be taken seriously by professional codebreakers. The *Gadsby* discussion will cause the light to dawn on some of the students who had not been able to figure out what the author of the mystery story in Set III did.

NEW LESSON

Transparency 9-9 is for possible use in introducing the new lesson in accord with the text.

Second day

One of the questions included in David Feldman's *When Do Fish Sleep? and other Imponderables of Everyday Life* (Harper and Row, 1989) is "Why are the commercials louder than the programming on television?"

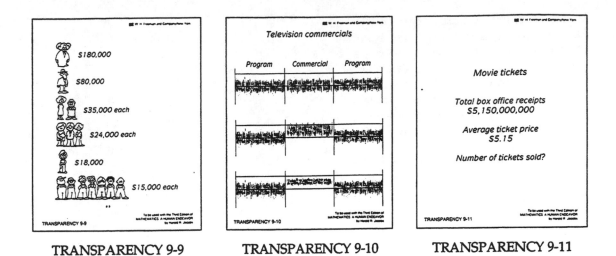

TRANSPARENCY 9-9   TRANSPARENCY 9-10   TRANSPARENCY 9-11

Show the top figure on Transparency 9-10, keeping the other figures covered. This figure represents the varying volume levels of a regular television program and a possible commercial. The lower line represents the average volume of the sound and the upper line represents the highest volume allowed by the Federal Communications Commission.

Show the second figure. Advertisers would like their commercials to be as loud as possible. If they merely turned up the volume, however, then the loudest parts of the commercial would be over the maximum volume allowed.

Show the third figure. To get around this, some advertisers use something called "volume compression" to reduce all of the sounds in the commercial to about the same volume. They can then raise the volume of the entire commercial without getting into trouble with the F.C.C. by going over the maximum legal volume. When this is done to a commercial, are its loudest parts any louder than the loudest parts of the regular program? (No.) How does the average volume of such a commercial compare with the average volume of the program surrounding it? (It is higher.)

ALTERNATE LESSON OPENER

The movie industry calculates the number of tickets sold in theaters in a surprising way: by dividing the total box office receipts by the average ticket price.

Show Transparency 9-11. In a recent year, the industry made an unexpected discovery.* The total box office receipts were $5,150,000,000 and, according to a government survey, it was thought that the average price for a ticket was $5.15. On the basis of this information, how many tickets do you think the movie industry assumed were sold? (One billion.)

Later, a more thorough survey of ticket prices was made by the National Association of Theater Owners and the Motion Picture Association of America and it was discovered that the average ticket price was not $5.15, but a dollar less. Using this information, they recalculated the number of tickets sold. What number did they get?

$$\left(\frac{\$5,150,000,000}{\$4.15} \approx 1,240,000,000. \right)$$

*Reported in the *Los Angeles Times*, March 9, 1994.

The theater owners' organization was pleased with this result because it not only showed that movie tickets were cheaper than previously thought but also that the movie theaters are drawing larger crowds than previously believed. One theater owner was quoted as saying that "this shows movies are a better value than a lot of people realize" and that he was not surprised that the average ticket price was lower than had been thought because many of his customers took advantage of bargain matinees and discount nights.

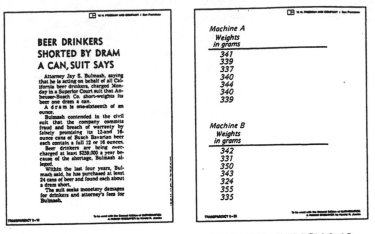

TRANSPARENCY 9-12   TRANSPARENCY 9-13

Measures of Variability

First day

Show Transparency 9-12, reading the article to your class. Does it seem reasonable that every 12-ounce can of beer manufactured by a company would contain *exactly* 12 ounces? (No. There are variations, even though they may be very small, in every manufacturing process.)

Show Transparency 9-13. (You may want to hand out duplicated copies of this transparency to save time.) Suppose that the contents of several cans of beer produced by two different machines are carefully weighed and found to be as shown in these lists. Copy the lists and find the average of the seven numbers in each. (It is 340 in each case; a weight of 340 grams is almost exactly equivalent to a weight of 12 ounces.)

Even though the average weights of the contents of cans produced by the two machines are the same, the machines do not seem to be giving the same results. Why? (The weights for machine B vary more widely than those for machine A.)

One way to show this difference is to give the difference between the two extremes in each set. For machine A, 344 − 337 = 7; for machine B, 355 − 324 = 31. The difference between the largest and the smallest numbers in a set of numbers is called its "range."

Another way to show how the numbers in a set vary is to figure out the "standard deviation." Although it is harder to figure out than the range, it is also more useful. Guide your students in finding the standard deviation in the weights for machine A and then give them a chance to try to determine the standard deviation for machine B. The results are shown at the top of the next page. (The numbers in both cases, as in the example and exercises of Lesson 4, have been chosen to prevent difficulties with fractions and square roots.)

149

Machine A:

Weights in grams	Difference from mean	Square of difference
341	1	1
339	1	1
337	3	9
340	0	0
344	4	16
340	0	0
339	1	+ 1
		28

Mean of squares: $\frac{28}{7}$ = 4.

Square root of mean: $\sqrt{4}$ = 2.

Machine B:

Weights in grams	Difference from mean	Square of difference
342	2	4
331	9	81
350	10	100
343	3	9
324	16	256
355	15	225
335	5	+ 25
		700

Mean of squares: $\frac{700}{7}$ = 100.

Square root of mean: $\sqrt{100}$ = 10.

Guide your students in using the mean and standard deviation to draw a scale (like that shown at the top of the next column) for the weights of the cans produced by machine A, pointing out that the mean weight, 340, is shown at the center and the standard deviation 2, is used to mark the units of the scale. Mark the seven weights with X's in the appropriate places along the scale.

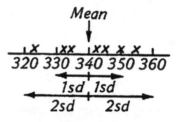

How many of the weights are 1 standard deviation or less from the mean? (Five.) What percentage of the seven weights is this? ($\frac{5}{7}$ × 100 ≈ 70%.) How many of the weights are 2 standard deviations or less from the mean? (All seven.) What percentage is this? (100%.)

Do the same for the weights of the cans of beer produced by machine B, drawing and marking a scale (as shown above) and asking the same questions. After your students have discovered that the answers are the same as before, you can point out that this is what makes the standard deviation such a useful number to know. There are many sets of data in which about 70% of the numbers are within 1 standard deviation of the mean and close to 100% of the numbers are within 2 standard deviations of it. (Our examples, containing so few numbers each, have been deliberately designed to conform to this pattern. It is probably best not to mention this to your students, however. The generalization applies to data that are "normally distributed.")

Show Transparency 9-14. This figure shows

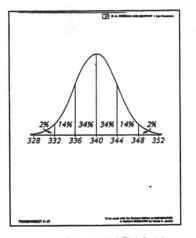

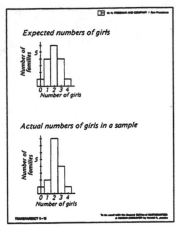

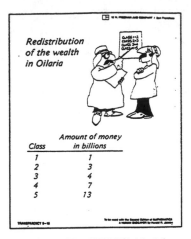

TRANSPARENCY 9-14   TRANSPARENCY 9-15   TRANSPARENCY 9-16

what the variation in weights in a large number of cans of beer might look like. You can see that the curve has a shape somewhat like that of a bell. It is called a "normal curve." The peak of the curve corresponds to the mean weight, which is 340 grams, or 12 ounces. The vertical lines separate the regions under the curve into slices corresponding to the standard deviation of the weights, which in this case is 4. What percentage of the weights are within 1 standard deviation of the mean? (68%.) In other words, 68% of the weights are between 336 and 344 grams. What percentage of the weights are within 2 standard deviations of the mean? (96%.) You can see from this that, although the mean tells us something only about the "typical" number in a set of numbers, the standard deviation tells us something about how the rest of the numbers are distributed around it.

Your students should now be ready to read the new lesson and begin the exercises.

REVIEW OF LESSON 3

Transparencies 9-15 and 9-16 are for use in discussing some of the exercises of Lesson 3.

Second day

A QUICK REVIEW OF THE NORMAL CURVE

Show Transparency 9-17. The telephone company has found that the times needed by operators to handle directory assistance calls are normally distributed as shown in this figure.*

What is the average time needed to handle one of these calls? (25 seconds.) What is the standard deviation in times? (5 seconds.) What percentage of directory assistance calls take less than 15 seconds to handle? (2%.) What percentage take more than 30 seconds? (16%.)

Remind your students that the percentages shown in this figure are the same for all normal curves: 68% of the numbers are within one standard deviation of the mean, 96% are within two standard deviations of it, and nearly 100% are within three standard deviations of it.

An especially good discussion of the nor-

---

*Based on information in *Introduction to the Practice of Statistics* by David S. Moore and George P. McCabe (W. H. Freeman and Company, 1989.)

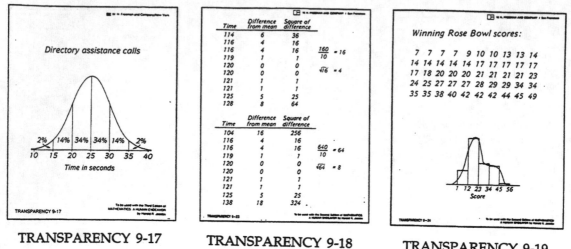

TRANSPARENCY 9-17   TRANSPARENCY 9-18   TRANSPARENCY 9-19

mal distribution is included in chapter 12 of *Lady Luck* by Warren Weaver (Anchor Books, 1963.)

Transparencies 9-18 and 9-19 are for use in discussing the Set I exercises of the lesson.

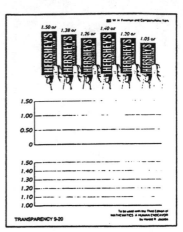

TRANSPARENCY 9-20

Displaying Data: Statistical Graphs

First day

Hand out duplicated copies of Transparency 9-20. Because of changes in the cost of its ingredients, the size of the Hershey bar changes from time to time.* This "bar" graph shows how the bar changed size in a six-year-period.

Graphs are often used to present statistical information because they can reveal patterns in a simple yet dramatic way. In the same way that a picture is said to be worth a thousand words, a graph might be said to be worth a thousand numbers. Unfortunately, however, graphs can also be drawn so as to deliberately distort the information that they supposedly represent.

Guide your students in drawing bars on the second figure on Transparency 9-22 to illustrate the relative weights of the Hershey bars in the first figure. Point out that this graph, drawn in a simpler form that the picture graph above it, is easy to understand because the lengths of the bars are drawn on a scale based on their weights. It is clear, for example, that the sixth bar is about two-thirds the size of the second bar.

Guide your students in drawing bars on the third figure to illustrate the same information in a different way. Point out that this graph represents the same changes in the size of the bars, yet it looks different because of the change in the vertical scale. The graph shows that the bar grew once and shrank four times, just as the previous graph does, but the changes are now exaggerated. Does the sixth bar look as though it is almost two-thirds the size of the second bar in this graph? (No. It looks one-tenth as large.)

With this introduction, your students should be ready to read Lesson 5 and begin the exercises.

*Stephen Jay Gould has written an amusing essay on this topic titled "Phyletic Size Decrease in Hershey Bars" which is included in his book *Hen's Teeth and Horse's Toes* (W. W. Norton and Company, 1983.)

153

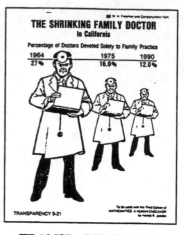

TRANSPARENCY 9-21

Second day

Show Transparency 9-21.* This picture graph appeared in a newspaper to illustrate the fact that there are less family doctors than there used to be.[†] According to the numbers beside the figures, in 1964 27% of the doctors in California were devoted solely to family practice. In 1990, the number had shrunk to 12%. Notice that 12% is less than half of 27% and that the 1990 doctor is less than half as tall as the 1964 one. (Illustrate this by standing the cut-out figure of the 1990 doctor at the left of the 1964 one so that their feet are aligned horizontally and then sliding the figure upward.)

Although the relative heights of the two doctors seem to be about right, there is some-

thing about them that is misleading. (Place the cut-out of the 1990 doctor so that its left arm aligns vertically with the left arm of the 1964 doctor. Slide the figure horizontally to the right.) Notice that the small doctor is not only about half as tall as the large one, but also about half as wide. (Slide the smaller figure into various positions on the larger one to show that its area is roughly one-fourth of the larger one.) If we compare the *areas* of the doctors rather than their heights, then their relative sizes are wrong. The comparison of areas exaggerates the change in the percentage of family doctors by suggesting that the number in 1990 is roughly *one-fourth* as many as in 1964, rather than approximately one-half.

There is something else that is misleading about the graph. (Draw vertical lines down the center of each doctor on the transparency and horizontal lines across them to serve as a time scale.) The time interval between the 1964 and 1975 doctors is 11 years and the time interval between the 1975 and 1990 doctors is 15 years. Does the horizontal spacing of the three doctors show this correctly? (No. The

---

*Make a second copy of the transparency so that you can cut out the figure of the 1990 doctor from it to use as described in this lesson plan.

[†]This graph is adapted from one that originally appeared in the *Los Angeles Times*, August 5, 1979, and is one of many good examples of misleading graphs included in the chapter titled "Graphical Integrity" in *The Visual Display of Quantitative Information* by Edward R. Tufte (Graphics Press, 1983.)

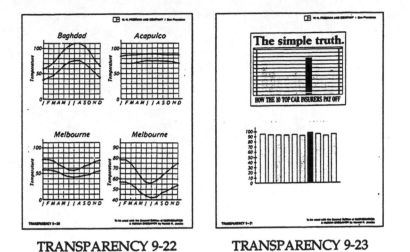

TRANSPARENCY 9-22   TRANSPARENCY 9-23

second time interval is larger so the third doctor should be farther to the right. Place the cut-out figure of the 1990 doctor over the 1990 doctor on the transparency and then slide it an appropriate distance to the right.) When the 1990 doctor is placed in the correct position, does the decrease in the percentage of family doctors seem as extreme? (No, because a curve through the tops of the three heads would not drop as steeply.)

These observations reveal that, although picture graphs seem like a nice way to compare data, they can actually be very misleading. In the case of the family doctor graph, the relative sizes of the three doctors are easily misinterpreted and their positions in the picture are actually incorrect.

Transparencies 9-22 and 9-23 are for use in discussing some of the Set I exercises of this lesson.

LESSON 6
Collecting Data: Sampling

First day

A basic part of statistics is the sampling process. To illustrate it, we will do a simple experiment. Hand out slips of paper and ask everyone to estimate, without any discussion, the total amount of money that the students in your school bring to school on a typical day. Have your students write down their names and guesses and then collect the slips. They should also record their guesses on their own paper.

How could we get a better idea of the answer to this question? (By finding how much money some of the students in the school have.) We will use our own class as a sample of the entire school. Ask everyone to determine the exact amount of money that he or she has at the moment and record this number on the paper. Now ask the students who have calculators with them to add the numbers as the students tell them. There will probably be a few people who have no money at all and a few who having surprisingly large amounts

with them. The total for the class usually turns out to be more than most of the students would have guessed. Ask your students to use this number to determine the average amount of money that each student in the class is carrying. (Tell your students the number of students present.) After the average has been checked, tell your students the typical number of students present in your school on a given day and ask them to make another estimate using this information. Then look through the slips of paper to see which students came closest to guessing this number.

Do you think that the estimate based on our sample is reasonably accurate? In discussing this, someone will surely point out that our sample may be too small or perhaps not typical because the amounts of money carried may vary systematically with students' interests, ages, or the classes that they take. Any systematic variation of this sort will bias the results.

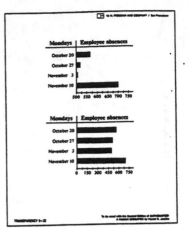

TRANSPARENCY 9-24

You can now explain the ideas of sample, population, size, and randomness in accord with the text. Be sure to point out that, for a sample of students in your school to be random, every student in the school has to have an equal chance of being selected. The size of the sample needed depends on the accuracy required; generally, a sample smaller than most people would expect will do the job.

You may wish to do some of the exercises in Lesson 6 orally rather than assigning them as written work. It is important to discuss examples in which data are biased in class so that your students will be primed to question the source of the data and the way in which it was gathered. Many sources of bias or bad data are hard to see. A few typical ones are mentioned in the answers for the exercises in this lesson.

For simplicity, only equal-probability, equally weighted random sampling is considered in this lesson. Statistical practice is more flexible. For example, in a survey to decide whether a new breakfast cereal will succeed in the market, it is important to weight people's responses proportionately to the amount of cereal they will be likely to purchase. Such samples are *weighted*. Similarly, a statistician may wish to make sure that a sample includes responses from people both in the east and in the west. This requires a *stratified* sample. Both procedures are departures from the equal-probability random sample stressed in the lesson, but they are designed to meet the same need, the need to obtain an unbiased representative sample.

Transparency 9-24 is for use in discussing some of the Set II exercises of Lesson 5.

Second day

Ann Landers once published a letter from a young couple asking whether having children was worth the problems entailed. She asked her readers: "If you had to do it over again, would you have children?" In the following weeks, she received nearly 10,000 responses, almost 70% of which said no.*

The people who wrote to Ann Landers were a sample of the population consisting of all parents in the United States. Do you think that the opinion of this sample accurately re-

---

*This example is from *Introduction to the Practice of Statistics* by David S. Moore and George P. McCabe (W. H. Freeman and Company, 1989).

flects the opinion of the population? The people who feel the most strongly tend to respond to such a poll. In this case, it was the people who were unhappy with their children. *Newsday* polled a *random* sample of parents on the same question and found that 91% were glad that they had had children. Another professional poll at about the same time got results agreeing with those obtained by *Newsday*.

In most classes, it is probably best to discuss all of the exercises that you have assigned as written work.

Review of Chapter 9

First day

In his book *Mathsemantics, Making Numbers Talk Sense* (Viking, 1994), Edward MacNeal writes: "The single most useful number for a marketing analyst in the United States is our total population. Not just for itself, but for the bearing it has on almost every other number."*

Without any preliminary class discussion, hand out slips of paper and ask your students to write down their name and estimate of the population of the United States. Collect the slips of paper and then, without divulging the answer, remark that it is likely that the esti-

---

*Chapter 14 of this book, titled "A Significant Number," contains an interesting discussion of people's answers in attempting to guess the population of the United States. MacNeal reports that in a group of 196 people, 71 had "no idea" and that the smallest answer given was 200,000, or about half the population of Staten Island. The largest number was 600,000,000,000, a figure more than one hundred times the population of the entire earth.

mates will vary considerably. Tell your students that you will read all of the responses to the class (without their names, of course), but that before you do this, you would like them to guess which one of the following will turn out to be closest to the correct answer: the *mean* of their answers, the *median*, or the *mode*.

Ask your students to record the responses as you read them and then give them time to try to determine all three numbers. This work will provide a good review of mean, median, and mode as well as the basis for a good discussion of why a particular one of the three might have been expected to have come closest. For example, which number is most affected by the largest guesses? (The mean.) Which number is most likely to be correct if several students were aware of the approximate population figure beforehand and did not have to guess it? (The mode.) Which number is least affected by the sizes of the worst guesses? (The median.)

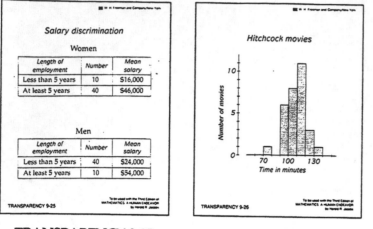

TRANSPARENCY 9-25   TRANSPARENCY 9-26

Second day

In an essay titled "'Readin', 'Ritin', and Statistics," Tim Robertson and Robert V. Hogg wrote: "Statistics now plays an important role in our everyday lives... One of the reasons that data are used so frequently . . . is that numerical information is often viewed as being objective and therefore beyond criticism. However, it is easy for numerical information to be very misleading."*

A shocking example of how numerical information can be interpreted in different ways is given on Transparency 9-25. (This example is adapted from one in the essay cited above.)

Suppose that a company has 100 employees and that their mean yearly salaries are as listed in the table shown. Does this company seem to discriminate in favor of men or women?

Find the total amount of money paid to the men during one year. (40 × $24,000 + 10 × $54,000 = $1,500,000.) What is the mean salary

paid to the men? $\left(\dfrac{\$1,500,000}{50} = \$30,000.\right)$ Find the total amount of money paid to the women during one year. (10 × $16,000 + 40 × $46,000 = $2,000,000.) What is the mean salary paid to the women? $\left(\dfrac{\$2,000,000}{50} = \$40,000.\right)$

On the basis of these results, against whom does the company seem to discriminate? Concerning these contradictory conclusions, Robertson and Hogg wrote:

"The subjective choice of which 'statistics' to report can change the message. Thus numerical evidence should be consumed not only with a certain amount of skepticism, but also with a considerable degree of skill. All of us should try to be critical consumers of statistics. Doublespeak exists in statistics as much as in the written or spoken word."

Transparency 9-26 is for use in discussing some of the Set II exercises of the review.

---

*This essay is included in the book *Mathematics Tomorrow*, edited by Lynn Arthur Steen (Springer-Verlag, 1981).

CHAPTER 10

Topics in Topology

Albert W. Tucker and Herbert S. Bailey, Jr. wrote in an article on topology in *Scientific American*:

> Topology is full of apparent paradoxes and impossibilities, and is probably more fun than any other branch of mathematics. The problems used as examples in this article may strike many readers as being mere curiosities. While the novel properties studied by topologists make the subject interesting and at times weird, this article will have failed in its purpose if it leaves the reader with the impression that topology is nothing more than a collection of Chinese puzzles. Topology is one of the most fundamental branches of mathematics.*

The final chapter of the text introduces the student to some simple problems of topological interest. The examples and applications included in it and in this guide are those that work well in the classroom. There are many others, some of which are quite technical. The existence of a critical instant at which a small shock can stop the heart has been explained on topological grounds.† Catastrophe theory, a comparatively recent branch of topology, has

many interesting applications. The physicist A. Zee wrote in his book *An Old Man's Toy, Gravity at Work and Play in Einstein's Universe* (MacMillan, 1989), "...from magnetic poles to cosmic strings, topology provides the common thread, underlying the methodological unity of physics."

The first lesson begins with the concept of topological equivalence and includes several illustrations of the Jordan Curve Theorem. Network theory is presented in three lessons, including Euler's analysis of the Königsberg-bridge problem, as well as other problems on traveling networks, and Euler's theorem relating the numbers of vertices, edges, and regions of a network. The exercises, in which entertaining puzzles are balanced with significant applications give the student a chance to discover a variety of patterns. The chapter ends with a series of interesting experiments with the Moebius strip and other surfaces.

---

*"Topology," January 1950.

†See *When Time Breaks Down*, by Arthur T. Winfree (Princeton University Press, 1987).

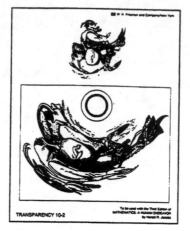

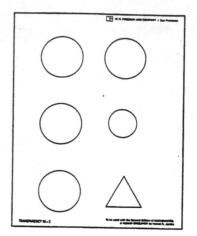

| TRANSPARENCY 10-1 | TRANSPARENCY 10-2 | TRANSPARENCY 10-3 |

LESSON 1

The Mathematics of Distortion

In about 1900, the McLoughlin Brothers Company in New York City published a beautiful set of twenty-four pictures to be viewed with a cylindrical mirror. The pictures, reproduced in full color, are available in the book, *The Magic Mirror: An Antique Optical Toy* (Dover Publications, 1979). The cover of the book is reproduced on Transparency 10-1. The picture illustrated on the cover, shown on Transparency 10-2, affords a nice way of introducing the first lesson on topology.

If possible, give each student a duplicated copy of Transparency 10-2 and a 5-by-8-inch piece of Mylar film.* The rectangle on the transparency represents one of the pages of the *Magic Mirror* book and the figure at the top shows its image as it appears in a cylindrical mirror. Your students can roll the film into such a mirror and then place its base on the circle to see for themselves how the image is formed.

---

*Mylar film is available in rolls from some art-supply and hobby stores.

After your students have done this, explain that the picture is an example of anamorphic art. An anamorphic picture is one in which the image of the object pictured has been stretched and distorted so that it is difficult to recognize what it represents. Such pictures must be viewed in a particular way, either from a certain perspective or with a suitable mirror, to see them without the distortion.

Although the picture looks very different in the distorted and undistorted forms, the two forms have several basic properties in common. The properties are "topological," and the branch of mathematics called "topology" deals with these properties. According to the article titled "General Topology" in the *Encyclopedia Britannica,*

Topology is the study of those properties that an object retains under deformation, specifically, bending, stretching and squeezing, but not breaking or tearing.

The figures on Transparency 10-3 can be

used to discuss the ideas of congruence, similarity, and topological equivalence. Explain that the two circles in the top row are called *congruent* in geometry because they have the same size and shape. Two figures that have the same shape but not necessarily the same size, such as the two circles in the second row, are said to be *similar*. Although the circle and triangle on the bottom row are neither congruent nor similar, they are alike in the sense that either one can be twisted and stretched into the shape of the other without connecting or disconnecting any points. For this reason, they are said to be *topologically equivalent*. In topology, triangles and circles are examples of *simple closed curves*. The ideas of topological equivalence and simple closed curves are explored in the exercises of this lesson.

In his book *The New Ambidextrous Universe* (W. H. Freeman and Company, 1990), Martin Gardner refers to research that reveals that very young children seem to grasp the idea of topological equivalence more readily than adults. He cites Jean Piaget and Bärbel Inhelder who, in their book *The Child's Conception of Space* (Humanities Press, 1956), "report on strong experimental evidence that children actually recognize topological properties be-fore they learn to recognize Euclidean properties of shape... When asked to copy a triangle, for example, very young children often draw a circle. The angles and sides of the triangle are less noticeable to them than the property of being a closed curve."

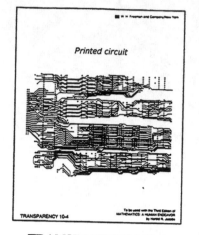

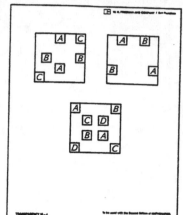

TRANSPARENCY 10-4   TRANSPARENCY 10-5

LESSON 2

The Seven Bridges of Königsberg: An Introduction to Networks

REVIEW OF LESSON 1

In the article "Chips for Advanced Computing" in the October 1987 issue of *Scientific American*, James D. Meindl wrote:

"Perhaps the most significant event of the past 50 years has been the emergence of a society whose chief product is information. The driving force of this transformation is the development of electronics technology, particularly the integrated circuit fabricated on a single chip of silicon. Since the advent of the integrated circuit in 1959, the number of transistors that can be squeezed onto a chip has increased from one to several million."*

The components of the first integrated cir-

cuit were hand-wired. Originally used primarily in military applications such as ballistic missiles, integrated circuits now appear in such widely ranging objects as calculators, microwave ovens, automobiles, Nintendo games, and running shoes. These circuits are made by a printing process which makes it possible to put an incredible number of components in a small space.

The figure on Transparency 10-4 shows part of a printed circuit.* The design of such a circuit is complicated by the fact that the connecting wires have to all lie in the same surface. As a result, they cannot cross each other.

Some comparatively simple circuit design problems were included in the Set II exercises of Lesson 1. Transparency 10-5 is for use in discussing them.

---

*Photographs of six increasingly complex chips made by the Fairchild Semiconductor Corporation from 1959 to 1985 are shown on page 80 of the October 1987 issue of *Scientific American*.

---

*This figure is from *Pearls in Graph Theory* by Nora Hartsfield and Gerhard Ringel (Academic Press, 1990).

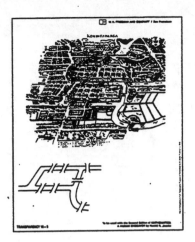

TRANSPARENCY 10-6

NEW LESSON

Transparency 10-6 can be used to introduce the new lesson. The transparency is more effective if you hand out duplicated copies of it so that your students can write directly on the figure below the map. Present the Königsberg-bridge problem to your students and let them make at least one attempt to solve it before continuing with the rest of the lesson. Euler's own account of his analysis of the problem is included in *Mathematics: An Introduction to Its Spirit and Use* (W. H. Freeman and Company, 1979). It also appears in the first volume of *The World of Mathematics*, edited by James R. Newman (Simon and Schuster, 1956), pages 573-580.

Leonhard Euler was a remarkable man; the most prolific mathematician in history, his writings fill more than 80 large volumes. Edna Kramer, in her book *The Nature and Growth of Modern Mathematics* (Hawthorn Books, 1970), says of him:

He was capable of remarkable concentration and would write his research papers with a child on each knee, while the rest of his youngsters raised uninhibited pandemonium all about him (he was the father of thirteen children in all).

Euler was totally blind during the last seventeen years of his life, yet he continued to do outstanding work until his death. A good account by Rudolph Langer of Euler's career is included on pages 213-218 of Dr. Kramer's book.

Building a house

	Activity	Time in days
A	Clear land	1
B	Build foundation	3
C	Build upper structure	15
D	Electrical work	9
E	Plumbing work	5
F	Complete exterior work	12
G	Complete interior work	10
H	Landscaping	6

TRANSPARENCY 10-7 · To be used with the Third Edition of *MATHEMATICS: A HUMAN ENDEAVOR* by Harold R. Jacobs

TRANSPARENCY 10-7

LESSON 3
Euler Paths

In his book *Graphs as Mathematical Models* (Prindle, Weber and Schmidt, 1977), Gary Chartrand wrote:

"The management sciences abound with situations requiring planning and scheduling. Many of the problems encountered are extremely complicated, and careful analysis of these problems commonly requires a sound background in a variety of mathematical subjects."

An example of such a planning and scheduling problem is given here and on Transparency 10-7. The example, building a house, is from Dr. Chartrand's book.

Given the times listed for the different activities, how long do you think it would take to build the house? Does it make sense to simply add the times and say 61 days? (No, because during part of the construction more than one activity might be occurring at the same time.) In a situation such as this, a network is helpful.

The activities can be represented by verti-

	Activity	Time in days
A	Clear land	1
B	Build foundation	3
C	Build upper structure	15
D	Electrical work	9
E	Plumbing work	5
F	Complete exterior work	12
G	Complete interior work	10
H	Landscaping	6

ces and the order of doing them by edges.

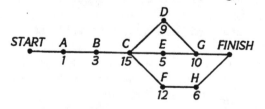

Guide your students in drawing a network such as the one above to represent the construction process. (Note that after the upper structure has been built, several activities can go on at the same time.)

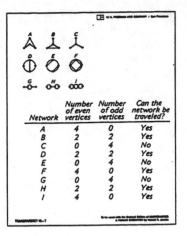

Network	Number of even vertices	Number of odd vertices	Can the network be traveled?
A	4	0	Yes
B	2	2	Yes
C	0	4	No
D	2	2	Yes
E	0	4	No
F	4	0	Yes
G	0	4	No
H	2	2	Yes
I	4	0	Yes

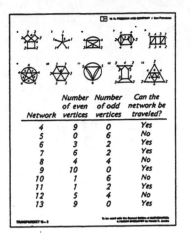

Network	Number of even vertices	Number of odd vertices	Can the network be traveled?
4	9	0	Yes
5	0	6	No
6	3	2	Yes
7	6	2	Yes
8	4	4	No
9	10	0	Yes
10	1	6	No
11	1	2	Yes
12	5	4	No
13	9	0	Yes

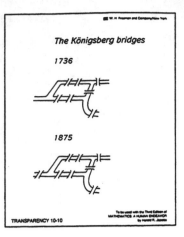

The Königsberg bridges

1736

1875

TRANSPARENCY 10-8 ·   TRANSPARENCY 10-9   TRANSPARENCY 10-10

Having drawn the network, consider the following questions. How many days from the beginning might the landscaping be finished? (37.) How many days from the beginning might the internal work be completed? (38.) In how many days might the house be completed? (38.)

REVIEW OF LESSON 2

Transparencies 10-8 and 10-9 are for use in discussing some of the exercises of Lesson 2.

NEW LESSON

Transparency 10-10 can be used to develop the new lesson in accord with the text.

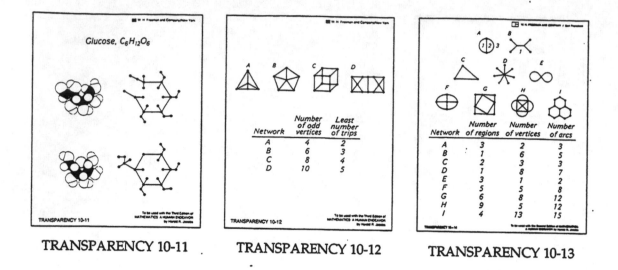

Glucose, $C_6H_{12}O_6$

Network	Number of odd vertices	Least number of trips
A	4	2
B	6	3
C	8	4
D	10	5

Network	Number of regions	Number of vertices	Number of arcs
A	3	2	3
B	1	6	5
C	2	3	3
D	1	8	7
E	3	1	2
F	5	5	8
G	6	8	12
H	9	5	12
I	4	13	15

TRANSPARENCY 10-11   TRANSPARENCY 10-12   TRANSPARENCY 10-13

LESSON 4

Trees

Show Transparency 10-11. Glucose is the basic fuel of biological cells. It is known by a variety of names, including dextrose, blood sugar and corn sugar.

Glucose molecules consist of carbon, hydrogen and oxygen atoms and exist in two different forms.* In the figures, each form is represented by a drawing of its atoms and by a network in which the vertices represent the atoms and the edges represent the bonds between them. Chemists refer to these networks as *structural formulas*.

The structural formulas reveal that one form of the glucose molecule contains a ring of atoms and that the other form does not. In mathematical terms, we might say that one network contains a simple closed curve and that the other does not. A network that does not contain any simple closed curves is called a *tree*.

Tree networks are sometimes measured by

---

*From Molecules by P. W. Atkins (Scientific American Library, 1987).

the largest number of edges that a path on them can have. In the case of the form of the glucose molecule that is a tree, the paths with the largest number of edges have 8 edges. (Show several such paths on the figure.) This number is called the *diameter* of the tree.

Because of the ring of atoms in the other form of the glucose molecule, a path on it can repeat edges (show this), so the idea of diameter is applied only to networks that are trees.

REVIEW OF LESSON 3

Transparencies 10-12, 10-13, 10-14, and 10-15 are for use in discussing some of the exercises of Lesson 3. The game invented by William Rowan Hamilton discussed in Set II is equivalent to finding a closed path along the edges of a dodecahedron that passes through every one of its corners exactly once. The figure in the margin of page 622 of the text is a plane projection of a dodecahedron.

Although the problems of finding an Euler path and a Hamilton path on a network seem

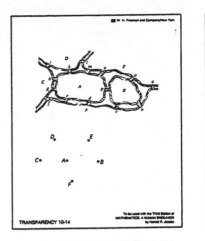

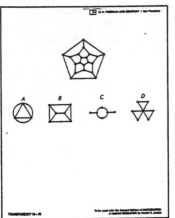

TRANSPARENCY 10-14   TRANSPARENCY 10-15

very similar, the second is much more difficult than the first. According to an article in *Scientific American*,*

"For Hamilton's problem . . . no efficient algorithm comparable to Euler's method has been found. The problem has been pondered by many of the best mathematicians of the past century, but the most efficient methods available today are fundamentally no better than exhaustive tabulation. On the other hand, all attempts to prove that there is no better method have also failed, and it must be considered a possibility that an efficient algorithm will be discovered tomorrow."

More information on the two problems can be found in the first two chapters of the COMAP book, *For All Practical Purposes* (W.H. Freeman and Company, 1993).

---

*"The Efficiency of Algorithms," by Harry R. Lewis and Christos H. Papadimitriou, *Scientific American*, January 1978.

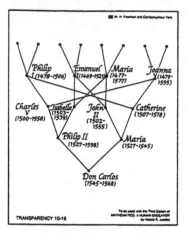

TRANSPARENCY 10-16

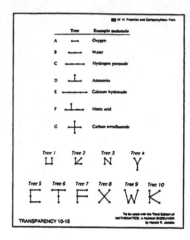

TRANSPARENCY 10-17

TRANSPARENCY 10-18

LESSON 5
The Moebius Strip and Other Surfaces

REVIEW OF LESSON 4

Show Transparency 10-16.* Don Carlos of Spain had an unusual family tree. To begin with, it is not a tree. Why not? (Because it contains some simple closed curves.) How many grandparents did Don Carlos have? (Four.) How many great-grandparents did he have? (Four.) Most people have eight great-grandparents. How did it come about that Don Carlos had only four? (His grandfather on his father's side, Charles V, was the brother of his grandmother on his mother's side, Catherine of Austria. His grandmother on his father's side, Isabella of Portugal, was the sister of his grandfather on his mother's side, John II.)

Although the family tree of Don Carlos seems very strange, each of us has a family tree with branches that are interconnected in this way if we look far enough back in time. If this

*This example is from *Visual Topology* by W. Lietzmann (London: Chatto and Windus, 1965).

were not so, the number of a person's ancestors in successive generations back in time would eventually become larger than the population of the entire earth. So, if you could trace your family tree for a sufficient number of generations, you would discover that it is not actually a tree in the topological sense.

Transparencies 10-17, 10-18, and 10-19 are for possible use in discussing some of the exercises of Lesson 4.

NEW LESSON

Show Transparency 10-20. One of the pioneers in topology was the mathematician and astronomer August Ferdinand Moebius.* Moebius spent most of his life in the city of Leipzig where he was a university professor

*The table on the transparency is adapted from *Moebius and His Band, Mathematics and Astronomy in Nineteenth-Century Germany*, edited by John Fauvel, Raymond Flood, and Robin Wilson (Oxford University Press, 1993).

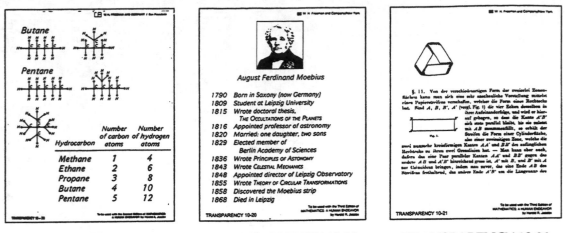

Below the transparency images:

TRANSPARENCY 10-19   **TRANSPARENCY 10-20**   **TRANSPARENCY 10-21**

and director of the university observatory. He wrote a number of books both on astronomy and mathematics, only some of which are listed in the table on the transparency. According to the *Encyclopedia Britannica*, Moebius is best known for his discovery of the Moebius strip.

Show Transparency 10-21. The figure at the top of the transparency is of a Moebius strip and is from a book written by Moebius soon after he discovered it. Below the figure is a description of how to make a Moebius strip, as it appeared in the same book. The directions say to take a strip of paper, shown as Figure 1, and tape the ends together so that the corner labeled A' falls on B and the corner B' falls on A. The result is a loop of paper with several strange properties. Following this brief introduction, your students should be ready to read Lesson 5 and begin the exercises.

Martin Gardner's *Mathematical Magic Show* (Alfred A. Knopf, 1977), contains a chapter with a wealth of information on the Moebius strip.

To do the experiments in the exercises, each student will need the following materials:

One sheet of graph paper, 4 units per inch (or 2 units per centimeter)

Scissors

Cellophane tape (one dispenser per two or three students)

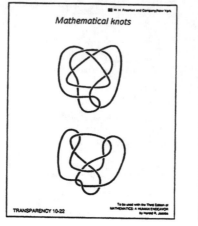

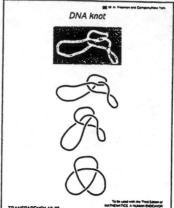

TRANSPARENCY 10-22   TRANSPARENCY 10-23

Review of Chapter 10

First day

Nearly everyone has had the experience of trying to untangle something that seems to be tied up in knots: perhaps fishing line or yarn or a garden hose. Sometimes the tangles come apart and other times no amount of pulling or twisting seems to help.*

One branch of topology is the theory of knots: the main problems of knot theory are to be able to tell whether something really is a knot, whether two knots are the same or different, and to identify and classify every possible type of knot.

Show Transparency 10-22. A *mathematical knot* is a simple closed curve in three-dimen-

---

*The information on knots in this lesson is based on the work of Ivars Peterson: "Knotty Problems" in *The Mathematical Tourist* (W. H. Freeman and Company, 1988), "Knot Physics" in *Islands of Truth* (W. H. Freeman and Company, 1990), and "Much Ado About Knotting" in *The Problems of Mathematics* (Oxford University Press, 1992).

sional space. The two curves shown in these figures are examples of mathematical knots. For 75 years these two knots were thought to be different, but in 1974, someone discovered that they are actually the same!

Show Transparency 10-23. The mathematician Ivars Peterson has written on the use of knot theory in science. The figure at the top of this transparency is a photograph taken through an electron microscope of a knotted loop of DNA. The figures below show that it can be untwisted to form a knot called the "trefoil" knot.

Peterson writes in *The Mathematical Tourist*:

"Molecular biologists are starting to use knot theory to understand the different conformations that DNA can take on. Recent advances in knot theory are helping them see how DNA becomes knotted or linked during replication and recombination and how the enzymes that do the cutting and gluing must perform their functions. Biologists can track the sequence of steps in which one structure is

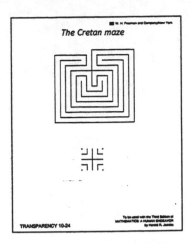

The Cretan maze

■ W. H. Freeman and Company/New York

To be used with the Third Edition of
MATHEMATICS: A HUMAN ENDEAVOR
by Harold R. Jacobs

TRANSPARENCY 10-24

TRANSPARENCY 10-24

gradually transformed into another during the basic, life-supporting processes that take place within cells . . .

Chemists interested in synthesizing new compounds are also beginning to pay attention to topology and knot theory. In general, they try to create new, unusual molecules by changing the way atoms are connected . . .

In 1981, David M. Walba of the University of Colorado managed to create a molecular Moebius strip with a half-twist . . . Walba's work, like that of molecular biologists working with the twists and turns of DNA, has itself suggested many new mathematical questions. In a crossover that enriches both pursuits, chemists and biologists are drawn into the fascinating mathematical world of knots and links as mathematicians tangle with molecules and chemistry."

The story of how David Walba and his colleagues synthesized the molecule in the shape of a Moebius strip is related in the chapter titled "The Moebius Molecule" in *Archimedes' Revenge, The Joy and Perils of Mathematics* by Paul Hoffman (W. W. Norton and Company, 1988).

Second day

Hand out duplicated copies of Transparency 10-24. According to an ancient legend, the Minotaur, a creature who was half-human and half-beast, was imprisoned in a maze on the island of Crete. The maze, also called a labyrinth, was an arrangement of turning corridors that seemed to have no beginning or end.

The maze on this transparency is known as the Cretan maze because of its association with this legend. It is the oldest maze known, appearing on a clay pot discovered in Syria and on a clay tablet discovered in Greece, both of which date back to about 1200 B.C. It also appeared on Cretan coins dating from 500 B.C. and on the walls of Pompeii.

In tracing a path through the maze, your students will discover that it has only one path, so that, unlike most modern mazes, you cannot get lost in it. Such a maze is called *unicursal*.

It is thought that the Cretan maze may have been created by starting with a nucleus consisting of a cross, four L's and four points. (This is shown in black on the grid below the maze on the transparency.) Starting with the

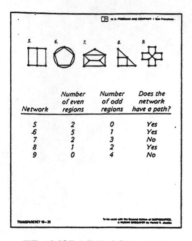

TRANSPARENCY 10-25

The table in the transparency reads:

Network	Number of even regions	Number of odd regions	Does the network have a path?
5	2	0	Yes
6	5	1	Yes
7	2	3	No
8	1	2	Yes
9	0	4	No

bottom of the cross and continuing from each successive point clockwise, each free endpoint (or point) is joined, going counterclockwise, to the next free endpoint (or point). The first two steps in this procedure are shown in the figures below. Your students can use this procedure, with the grid as a guide for drawing the lines, to construct the entire maze.

Maze patterns were a popular way of decorating floors in ancient Rome. According to Anthony Phillips, a mathematician at the State University of New York at Stony Brook, the Roman mazes occur in 25 different topological types. In an article on the subject, he states:

"Devising a unicursal maze to fill out an area in an interesting and symmetrical way is a topological problem as well as an aesthetic problem. Given constraints on size and overall organization, there is only a small number of topologically distinct solutions ... Each of

the solutions that occur was first discovered by someone, somewhere; a fascinating element of the study of ancient mazes is the contact with these ingenious, unsung topologists of the past."*

Transparency 10-25 is for possible use in discussing some of the Set II exercises of the review lesson.

_____

*See "The Topology of Roman Mosaic Mazes" by Anthony Phillips in *The Visual Mind: Art and Mathematics*, edited by Michele Emmer (MIT Press, 1993). The nucleus method for drawing the Cretan maze is taken from this article.

Answers to Exercises

Chapter 1, Lesson 1 (pp. 8-11)

Set I

1.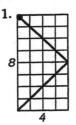

2.

3.

4.

5.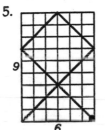

6.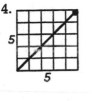

7.

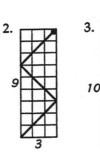

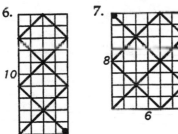

8.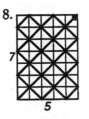

9. Table 4.

10. It is square (or, it has equal dimensions).

11. Tables 1, 2, 3, and 4.

12. Yes.

13. No. It doesn't seem as if it can end up in the same corner from which it started.

14. Table 8.

15. (Drawing of any table whose dimensions are relatively prime.)

Set II

1.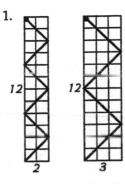

2.

Dimensions of table	Number of segments in path
12 by 1	12
12 by 2	6
12 by 3	4
12 by 4	3

3. By dividing the length by the width.

4.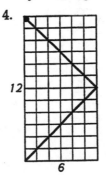

5. Tables 1, 2, 3, and 4.

6. $\frac{9}{3} = \frac{6}{2} = 3$.

7.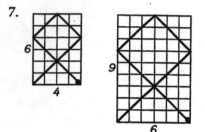

8. The paths have the same shape.

9.

3 by 2.

10.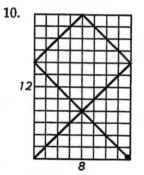

11.

Dimensions of table	Corner ball ends up in
2 by 1	Upper left
3 by 2	Lower right
4 by 3	Upper left
5 by 4	Lower right
6 by 5	Upper left

12. 8 by 7.

13.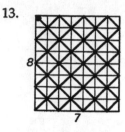

14.

Dimensions of table	Corner ball ends up in
100 by 99	Upper left
101 by 100	Lower right

Set III

1. If the ball ends up in the upper left corner, the reflection in the mirror is the same as the actual path.

2. If the ball ends up in the lower right corner, the reflection in a vertical mirror is the same as the actual path.

3. If the ball ends up in the upper right corner, the path looks the same rightside up and upside down.

Chapter 1, Lesson 2 (pp. 14-18)

Set I

1.

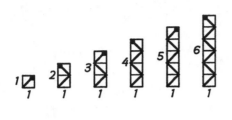

2. The tables with lengths 2, 4, and 6.

3. The tables with lengths 1, 3, and 5.

4. Yes.

5. On a table whose length is even and whose width is 1, the ball ends up in the upper left corner.

6. (Examples.)

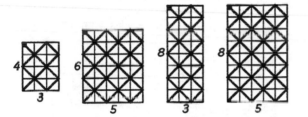

7. On a table whose length is even and whose width is odd, the ball ends up in the upper left corner.

8. (Examples.)

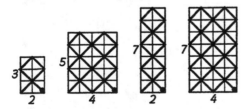

9. On a table whose length is odd and whose width is even, the ball ends up in the lower right corner.

10. (Examples.)

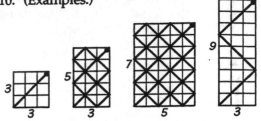

11. On a table whose length and width are both odd, the ball ends up in the upper right corner.

Set II

1. No.

2.

3. Yes.

4. $\frac{7}{2}$.

5.

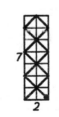

6. Yes. On a table whose length is odd and whose width is even, the ball ends up in the lower right corner.

7. $\frac{6}{5}$.

8.

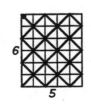

9. Yes. On a table whose length is even and whose width is odd, the ball ends up in the upper left corner.

10. Reduce the ratio of the dimensions to lowest terms.

11. $\frac{95}{85} = \frac{19}{17}$. Upper right because reduced length and width are both odd.

12. $\frac{105}{100} = \frac{21}{20}$. Lower right because reduced length is odd and width is even.

13. $\frac{120}{110} = \frac{12}{11}$. Upper left because reduced length is even and width is odd.

4. 191. (The ratio $\frac{96}{95}$ cannot be reduced; $96 + 95 = 191$.)

5. 21. $\left(\frac{110}{100} = \frac{11}{10}; \ 11 + 10 = 21.\right)$

6. 13. $\left(\frac{105}{90} = \frac{7}{6}; \ 7 + 6 = 13.\right)$

Set III

1.

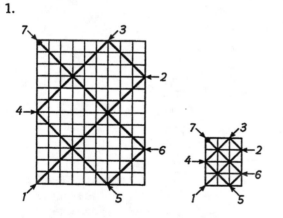

2. The tables are similar because $\frac{12}{9} = \frac{4}{3}$, and so the paths on the tables have the same shape.

3. Reduce the ratio of the dimensions of the table to lowest terms. The number of hits can be found by adding the reduced dimensions.

Chapter 1, Lesson 3 (pp. 20-24)

Set I

1. $1 + 3 + 5 + 7 + 9 = 5 \times 5$

2. Yes.

3. $1 + 3 + 5 + 7 + 9 + 11 + 13 = 7 \times 7$

4. Yes.

5. $1 + 3 + 5 + 7 + 9 + 11 + 13 + 15 + 17 + 19 = 10 \times 10$

6. Yes.

7. Yes.

8. Inductive.

9. From 7×2.

10. From 7×9.

11. From 4×2.

12.
```
      53
   ×  86
   ─────
      30
    4018
   +  24
   ─────
    4558
```

13.
```
      75
   ×  75
   ─────
      35
    4925
   +  35
   ─────
    5625
```

14.
```
      21
   ×  34
   ─────
      08
    0604
   +  03
   ─────
    0714
```

15. Inductive.

Set II

1. The length of the pendulum seems to be equal to the time of the swing multiplied by itself.

2. 25 units.

3. 100 units.

4. The pressure gets smaller.

5. The pressure is divided by 2.

6. The pressure is divided by 4.

7. The pressure would be divided by 3.

8. 40 units.

9. 7.5 units.

10. $24 + 4 = 28$

11. $192 + 4 = 196$

12. That there was a undiscovered planet between Mars and Jupiter.

Set III

1. ✚

2. ∴ (The figure looks the same upside-down.)

3. ✚ or ★.

4. ✚ (Starting at the upper left corner and spiraling clockwise, there is first one symbol, then two symbols, then three, then four, then five, then six, then seven, then eight.)

5. (Student answer.)

Chapter 1, Lesson 4 (pp. 27-30)

Set I

1. They are four years apart.
2. Inductive.
3. 1916.
4. (Student answer. There were no Olympics in 1916 because of World War I.)
5. 11
6. 111
7. 1,111
8. 11,111
9. $12,345 \times 9 + 6 = 111,111$
10. It is true.
11. $123,456,789 \times 9 + 10 = 1,111,111,111$
12. It is true.
13. No.
14. 333,333,331
15. Inductive.
16. 19,607,843
17. That it is not prime.
18. That the results obtained by inductive reasoning are not always correct.

Set II

1. 16 and 32.
2.

5 points

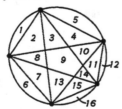

6 points

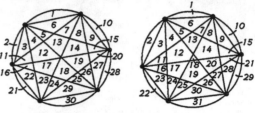

3. 16 and 31 (or, 30)
4. No.
5. The top of the T has $2 \times 9 = 18$ squares. The bottom has $6 \times 3 = 18$ squares. $18 + 18 = 36$.
6. 63 square units.
7.

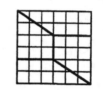

8. 36 square units.
9.

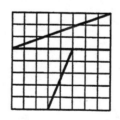

10. 36 square units.
11. No.
12.

13. 64 square units.
14.

15. 65 square units.

16. It seems to change.

17. (This is a hard question.) The explanation is that the pieces do not actually fit perfectly together. The exaggerated drawing of the rectangle below illustrates this.

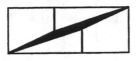

More information on geometrical vanishes can be found in *Mathematics, Magic and Mystery* by Martin Gardner (Dover Publications, 1956).

Set III

1. The digits form a pattern in sets of threes:

     004 115 226 337

2. 448 559 66

3.
$$243 \overline{)\,1.000\,000\,000\,000\,000\,000\,00}$$
     quotient: 0.004 115 226 337 448 559 67

4. No. The pattern breaks down at the 20th decimal place.

Chapter 1, Lesson 5 (pp. 33-38)

Set I

1. If the pennies are placed on squares of the same color, A wins; if they are placed on squares of different colors, B wins.

2. Inductive.

3. No.

4. One is brown and one is white.

5. Seven of the squares must be brown and seven of the squares must be white.

6. Six are white and eight are brown.

7. Six are brown and eight are white.

8. Seven are brown and seven are white.

9. Deductive.

10. It would have to be white because the label is *wrong*.

11. Yes. Yes.

12. The other marble would have to be red because the label is wrong.

13. No. (It could be red or white.)

14. The other marble would have to be red.

15. The other marble would have to be white.

16. The box "1 red, 1 white."

17. It must be red.

18. One must be red and one must be white.

19. Both must be white.

Set II

1.
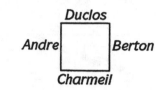

2. Andre is the butcher, Berton the baker, Charmeil the grocer, and Duclos the tobacconist.

3. 8.

4. 12.

5. 6.

6. 125. $(5 \times 5 \times 5 = 125.)$

7. Three.

8. 8.

9. Two.

10. 36. $(12 \times 3 = 36.)$

11. One.

12. 54. $(6 \times 9 = 54.)$

13. Inside the large cube.

14. 27. $(3 \times 3 \times 3 = 27.)$

15. $8 + 36 + 54 + 27 = 125.$

16. 1,728. $(12 \times 12 \times 12 = 1728.)$

17. 8.

18. 120. $(12 \times 10 = 120.)$

19. 600. $(6 \times 100 = 600.)$

20. 1,000. $(10 \times 10 \times 10 = 1,000.)$

21. $8 + 120 + 600 + 1,000 = 1,728.$

22. The year 2000 begins on Saturday. Dividing 7 into 365 shows that a year contains 52 weeks and 1 day.

23. The year 2001 begins on Monday. A leap year contains 52 weeks and 2 days.

24. 14. There are 7 calendars for years containing 365 days, one starting with each day of the week. There are 7 more calendars for leap years.

Set III

1. The hypotenuse.

2. The legs.

3. The area filled by the large square must be equal to the area filled by the two small squares put together.

Chapter 1, Lesson 6 (pp. 41-44)

Set I

1. □□

2. □ ○○

3. □ ○

4. □□□

5. □ ○

6. □□□□ ○○○

7. $4n + 2$.

8. $n + 5$.

9. 5.

10. $5n$.

11. Choose a number. □ n

    Add 3.   □ ○○○ $n + 3$

    Multiply by 2.   □□ ○○○○○○ $2n + 6$

    Add 4.   □□ ○○○○○○○○○○ $2n + 10$

    Divide by 2.   □ ○○○○○ $n + 5$

    Subtract the number first thought of.   ○○○○○   5

    The result is 5.

12. Choose a number. □ n

    Double it.   □□ $2n$

    Add 9.   □□ ○○○○○○○○○ $2n + 9$

    Add the number first thought of.   □□□ ○○○○○○○○○ $3n + 9$

    Divide by 3.   □ ○○○ $n + 3$

    Add 4.   □ ○○○○○○○ $n + 7$

    Subtract the number first thought of.   ○○○○○○○   7

    The result is 7.

13. Choose a number. □ n

    Triple it.   □□□ $3n$

    Add the number one larger than the number first thought of.   □□□□ ○ $4n + 1$

    Add 11.   □□□□ ○○○○○○○○○○○ $4n + 12$

    Divide by 4.   □ ○○○ $n + 3$

    Subtract 3.   □ n

    The result is the original number.

Set II

1. (Example.)
   896
   $896 \times 7 = 6{,}272$
   $6{,}272 \times 11 = 68{,}992$
   $68{,}992 \times 13 = 896{,}896$

2. It contains the original three-digit number written twice.

3. Inductive reasoning.

4. 1,001.

5. (Example.)

$$\begin{array}{r} 896 \\ \times\ 1,001 \\ \hline 896 \\ 896 \\ \hline 896,896 \end{array}$$

6. 1,001.

7. Yes.

8. (Example.)
   92
   $92 \times 13 = 1,196$
   $1,196 \times 21 = 25,116$
   $25,116 \times 37 = 929,292$

9. (A second example.)

10. The result is a six-digit number consisting of the original two-digit number written three times.

11. 10,101

12. Multiplying a number by 13, 21, and 37 in succession gives the same result as multiplying it by 10,101.

Set III

1. (Student answer.)

2. (Student answer.)

3. The result is twice the current year.

4. Adding the year of your birth and your age as of the last day of the current year gives the current year. Adding the year in which an important event in your life occurred and the number of years since that event occurred gives the current year. Adding all four numbers, therefore, gives twice the current year.

Chapter 1, Review (pp. 45-49)

Set I

1. $16 = 3 + 13$ (other answers are possible for all of these)
   $18 = 5 + 13$
   $20 = 3 + 17$
   $22 = 3 + 19$
   $24 = 5 + 19$
   $26 = 3 + 23$
   $28 = 5 + 23$
   $30 = 7 + 23$
   $32 = 3 + 29$
   $34 = 3 + 31$

2. No. (No one has so far!)

3. Inductive.

4.

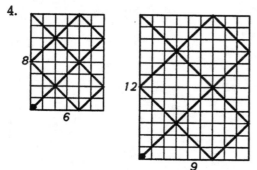

5. They are similar (or, the same).

6. The tables have the same shape because

$$\frac{8}{6} = \frac{12}{9} = \frac{4}{3}.$$

7. 13.

8. 10.

9. Inductive.

10. Deductive.

11. Because Tom is the only one who can always be believed.

12. Tom must be the golfer on the right because he would not say what the other golfers are saying.

13. From what Tom says. (Harry is in the middle.)

14. He is lying.

Set II

1. 9,801.

2. 998,001.

3. 99,980,001.

4. 9,999,800,001.

5. Inductive.

6.

Piece	Squares it would cover
A	2 brown and 2 white
B	2 brown and 2 white
C	3 brown and 1 white (or 1 brown and 3 white)
D	2 brown and 2 white
E	2 brown and 2 white

7. No. The board contains 10 brown and 10 white squares. The pieces would cover either 9 brown and 11 white squares or 11 brown and 9 white squares.

8. (Example.)

$$\begin{array}{r} 783 \\ -\ 387 \\ \hline 396 \end{array} \qquad \begin{array}{r} 396 \\ +\ 693 \\ \hline 1{,}089 \end{array}$$

9. The final results in both examples are the same number: 1,089.

10. No.

11. (Example.)

$$\begin{array}{r} 594 \\ -\ 495 \\ \hline 99 \end{array} \qquad \begin{array}{r} 99 \\ +\ 99 \\ \hline 198 \end{array}$$

12. The final results are the same number: 198.

13. The result would be 0.

14. Miss Blue is wearing a green dress, Miss Black is wearing a blue dress, and Miss Green is wearing a black dress.

Set III

1. 12.

2. No.

3. 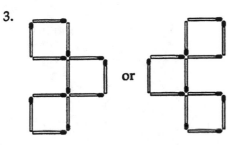 or

4. 2.

Chapter 1, Further Exploration

Lesson 1 (pp. 50-51)

1. a) (Examples.)

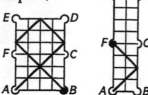

b) The ball cannot end up in pocket A or pocket E. Examples suggest that pocket C always lies in the path of any ball that might go to pocket E, so that the ball ends up in pocket C instead.

2. a)

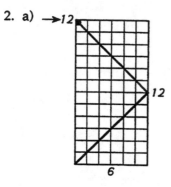

b)

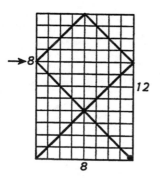

c)

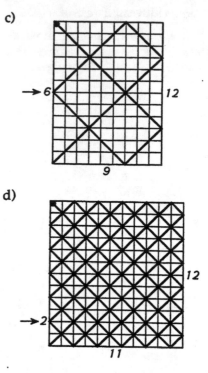

d)

e) It is one-half the number of the closest lighted point on mirror A.

Lesson 2 (p. 51)

The drawing below shows that a ball can travel in an endless loop on a table that is not square.

Paths that are loops are possible only on tables whose dimensions have a common factor.

Lesson 3 (pp. 52-53)

1. a) The sums of the three numbers in each row, each column, and both diagonals is the same: 45.

b) Each number is the number of letters in the name of the corresponding number in the figure on the left.

c) The sums of the three numbers in each row, each column, and both diagonals is the same: 21. Also, the square contains the nine consecutive numbers from 3 through 11.

2. a and c)

Ending decimals	Non-ending decimals
$\dfrac{1}{2}, \dfrac{1}{4}, \dfrac{3}{4}, \dfrac{1}{5}, \dfrac{2}{5},$	$\dfrac{1}{3}, \dfrac{2}{3}, \dfrac{1}{6}, \dfrac{5}{6}, \dfrac{1}{7},$
$\dfrac{3}{5}, \dfrac{4}{5}, \dfrac{1}{8}, \dfrac{3}{8}, \dfrac{5}{8},$	$\dfrac{2}{7}, \dfrac{3}{7}, \dfrac{4}{7}, \dfrac{5}{7}, \dfrac{6}{7},$
$\dfrac{7}{8}, \dfrac{1}{10}, \dfrac{3}{10}, \dfrac{7}{10},$	$\dfrac{1}{9}, \dfrac{2}{9}, \dfrac{4}{9}, \dfrac{5}{9}, \dfrac{7}{9},$
$\dfrac{9}{10}, \dfrac{1}{16}, \dfrac{1}{20}$	$\dfrac{8}{9}, \dfrac{1}{11}, \dfrac{1}{12}, \dfrac{1}{13},$
	$\dfrac{1}{14}, \dfrac{1}{15}, \dfrac{1}{17}, \dfrac{1}{18}, \dfrac{1}{19}$

b) The denominator.

d) The decimal form ends if the denominator has factors of only 2's and 5's.

Lesson 4 (pp. 53-54)

1. a) None of the lines disappear.

b) The rearrangement changes the number of lines in part because two of the lines of the original poem are entirely on one side or the other of the vertical line. The other six lines are broken into two parts. These parts are redistributed, together with the two unbroken lines, to form the new poem with seven lines. The lines of the new poem are slightly longer than those of the original poem.

2. a) 11 $1 \times 2 \times 3 \times 4 \times 5 \times 6 \times 7 \times 8 \times$
   $9 \times 10 + 1 = 3{,}628{,}801$
   Is 11 a factor of 3,628,801?   Yes.

b) If the number is prime, the answer is yes. More specifically, if n is prime, n is a factor of $1 \times 2 \times 3 \times 4 \times \cdots \times (n-1) + 1$.

Lesson 5 (pp. 54-56)

1. a) | | ⊓

b) Each vertical bar represents a line of the poem. The connected bars represent the lines that rhyme.

c) Symbol 8.

d) Symbol 6.

e) Symbol 10.

f) Symbol 15.

g) Five. Representing the lines that rhyme by the same letter, the patterns are: aaa, aab, aba, abb, abc.

h) ⊓ ⊓| ⊓ |⊓ |||

2. Each cube must have a 1 and a 2 in order to make the dates 11 and 22. Each cube must also have a 0 to make all of the dates 01, 02, 03, 04, 05, 06, 07, 08, and 09 possible. (If only one cube had a 0, the other would have to have nine digits.)

The digits on the cube on the right, therefore, are

0, 1, 2, 3, 4, and 5.

The remaining digits 6, 7, 8, and 9, must be on the cube on the left. Because it also must have a 0, 1, and 2, the face numbered 6 will have to be turned upside down when a 9 is needed. So the digits on the cube on the left are

0, 1, 2, 6/9, 7, and 8.

Lesson 6 (pp. 56-57)

1. a) (Student table.)

   b) The final sum is 7 more than the sum of the three numbers showing at the end.

   c) The numbers on a die are arranged so that numbers on opposite faces total 7. Consequently, the two unknown numbers in the final sum add up to 7.

2. a) Regardless of which brown card is chosen at the beginning, the coin ends up on the ace of diamonds.

   b) When the coin is first moved left or right to the nearest black card, there are only four cards on which it can end up: the three of spades, the ten of spades, the four of clubs, or the queen of spades. When the coin is then moved vertically up or down to the nearest brown card, it will land on either the nine of hearts or the king of diamonds. The next move, diagonally to the nearest black card, carries it to either the four of clubs or the two of clubs. The final move, down or to the right to the nearest brown card, carries the coin to the ace of diamonds.

Chapter 2, Lesson 1 (pp. 61-66)

Set I

1. Yes. Each successive term is found by adding the same number, 2.

2. The odd numbers.

3. The even numbers.

4. 2, 5, 8.

5. 3.

6. 11, 14, 17.

7. 32.

8. 176.

9. That the numbers continue.

10. 2, 9, 16, 23, 30.

11. 7.

12. 2, 7, 12, 17, $\underline{22}$, $\underline{27}$

13. 5, 13, 21, $\underline{29}$, $\underline{37}$

14. 11, 15, $\underline{19}$, 23, $\underline{27}$

15. $\underline{2}$, $\underline{11}$, 20, 29, 38

16. 4, $\underline{11}$, 18, $\underline{25}$, 32

17. $\underline{17}$, 33, $\underline{49}$, 65, $\underline{81}$

18. 3.

19. -3.

20. 27 cm.

21. 24, 21, 18, 15, $\underline{12}$, $\underline{9}$, $\underline{6}$, $\underline{3}$, $\underline{0}$

22. 9.

23. 10, $\underline{40}$, 70

24. 10, $\underline{30}$, $\underline{50}$, 70

25. 10, $\underline{25}$, $\underline{40}$, $\underline{55}$, 70

26. 10, $\underline{22}$, $\underline{34}$, $\underline{46}$, $\underline{58}$, 70

27. 10, $\underline{20}$, $\underline{30}$, $\underline{40}$, $\underline{50}$, $\underline{60}$, 70

28. 2. $\left(\dfrac{28-16}{6}\right)$

29. 16, $\underline{18}$, $\underline{20}$, $\underline{22}$, $\underline{24}$, $\underline{26}$, 28

30. 34.

31. 40. $\left(\dfrac{400-240}{4}\right)$

32. 240, $\underline{280}$, $\underline{320}$, $\underline{360}$, 400

Set II

1. 31, 38, 45, 52, 59, $\underline{66}$.

2. $3 + 7(9) = 3 + 63 = \underline{66}$.

3. 27, 31, 35, 39, 43, 47, 51, 55, 59, 63, 67, 71, 75, 79, 83, $\underline{87}$.

4. $11 + 4(19) = 11 + 76 = \underline{87}$.

5. $8 + 7(10) = 8 + 70 = \underline{78}$.

6. $6 + 4(24) = 6 + 96 = \underline{102}$.

7. $100 + (-3)(15) = 100 - 45 = \underline{55}$.

8. $55 - 3 = \underline{52}$.

9. $24 + 11(39) = 24 + 429 = \underline{453}$.

10. $5 + 25(70) = 5 + 1,750 = \underline{1,755}$.

11. $1,755 + 25 = \underline{1,780}$.

12. $t_1 + 3d$, $t_1 + 4d$.

13. $t_1 + 9d$.

14. $t_1 + (n-1)d$.

Set III

1. $\dfrac{16 \cdot 17}{2} = 136$.

2. 136.

3. 144.

4. $\dfrac{12 \cdot 24}{2} = 144$.

5. 210. $\left(\dfrac{20 \cdot 21}{2}\right)$

6. 5,050. $\left(\dfrac{100 \cdot 101}{2}\right)$

Chapter 2, Lesson 2 (pp. 69-74)

Set I

1. By multiplying by 3.
2. By multiplying by 4.
3. Geometric.
4. 3, 6, 12.
5. 2.
6. 24, 48, 96.
7. 4, 12, 36, <u>108</u>, <u>324</u>
8. 11, 121, <u>1,331</u>, <u>14,641</u>
9. 5, 20, <u>80</u>, 320, <u>1,280</u>
10. <u>18</u>, <u>36</u>, 72, 144, 288
11. 0, <u>0</u>, <u>0</u>, <u>0</u>
12. 16, 40, 100, <u>250</u>, <u>625</u>
13. 81, 54, <u>36</u>, <u>24</u>, <u>16</u>
14. Arithmetic; difference is 5.
15. Geometric; ratio is 4.
16. Neither.
17. Geometric; ratio is 0.5.
18. Arithmetic; difference is 17.
19. Neither.
20. Arithmetic; difference is -9.
21. Geometric; ratio is 1.5.
22. 2.
23. 27.5, 55, 110, 220, 1,760, 3,520.
24. 27.5, ~~55~~, 110, ~~220~~, 440, ~~880~~, 1,760, ~~3,520~~;
    Yes.
25. 0.75.
26. 540 cm and 405 cm.

Set II

1. 10.
2. Six. (They are 1, 10, 100, 1,000, 10,000, and 100,000.)
3. 5, 50, 500, 5,000.
4. 50.
5. 5, 10, 20.
6. 2, 10, 50 or 20, 100, 5,000.
7. 1, 100, 10,000 or 10, 1,000, 100,000.
8. $2 + 3 \cdot 3$ and $2 + 3 \cdot 4$.
9. $2 \cdot 3^3$ and $2 \cdot 3^4$.
10. $2 + 3 \cdot 9$.
11. 29.
12. $2 \cdot 3^9$.
13. 39,366.
14. 6, 24, 96, 384, 1,536.
15. $6 \cdot 4^{99}$.
16. $7 \cdot 5^{10}$.
17. $2 \cdot 10^{49}$.
18. $3 \cdot 8^{122}$.
19. $t_1 \cdot r^7$.
20. $t_1 \cdot r^{n-1}$.

Set III

1. $8,000. ($16 \times \$500$.)
2. 32.
3. 64, 128, and 256.
4. 511. ($1 + 2 + 4 + 8 + 16 + 32 + 64 + 128 + 256$.)
5. 31. (The people on levels 1 through 5.)
6. $248,000. ($31 \times \$8,000$.)
7. The people on level 8 (the last level).
8. Infinitely many.

Chapter 2, Lesson 3 (pp. 77-82)

Set I

1. 1, 2, 4, 8, 16.
2. The binary sequence.
3. 32, 64, 128, 256, 512.
4. $4 + 1$.
5. $16 + 4$.
6. $32 + 8 + 2$.
7. $64 + 4 + 2 + 1$.
8. $64 + 16 + 8 + 4 + 2 + 1$.

	Number	Binary numeral						
		64	32	16	8	4	2	1
	19			1	0	0	1	1
	100	1	1	0	0	1	0	0
9.	5					1	0	1
10.	20			1	0	1	0	0
11.	42			1	0	1	0	1	0
12.	71	1	0	0	0	1	1	1
13.	95	1	0	1	1	1	1	1

14. 15. $(8 + 4 + 2 + 1.)$
15. 31. $(16 + 8 + 4 + 2 + 1.)$
16. 63. $(32 + 16 + 8 + 4 + 2 + 1.)$
17. 20. $(16 + 4.)$
18. 40. $(32 + 8.)$
19. 105. $(64 + 32 + 8 + 1.)$
20. 210. $(128 + 64 + 16 + 2.)$
21. Multiplying the number by 2.
22. The letters A through C become 0, 1, and 2 and the letters E through Z become 4 through 23.
23. 32. (The letter groups correspond to the numbers 0 and 1 through 31.)

Set II

1. 2 4 8 16 32 64
2. 1 3 7 15 31 63
3. They are each 1 less.
4.

t_1	t_2	t_3	t_4	t_5	t_6	t_7
1	2	4	8	16	32	64

t_8	t_9	t_{10}	t_{11}	t_{12}
128	256	512	1,024	2,048

5. 127.
6. $128 - 1 = 127$.
7. 1,023.
8. The 65th term.
9. 18,446,744,073,709,551,615.
10. $31.
11. $32.
12. $1 ahead. ($32 − $31.)
13. $127.
14. $128.
15. $1 ahead. ($128 − $127.)
16. $18,446,744,073,709,551,615.
17. $18,446,744,073,709,551,616.
18. $1 ahead.
19. $36,893,488,147,419,103,231 behind.

Set III

1. If the hexagrams are turned sideways with the bottom bar at the left and the solid lines replaced with 0's and the broken lines with 1's, they become the binary numerals they represent.

2. Because six solid lines correspond to "000000," or 0, and six broken lines correspond to "111111," or 63, there are $1 + 63 = 64$ hexagrams possible in all.

3.

K'uei	Chia jēn	I	Sung	Lū
10	20	30	40	50

Chapter 2, Lesson 4 (pp. 84-91)

Set I

1. 1, 4, 9, 16, 25, 36, <u>49</u>, <u>64</u>, <u>81</u>, <u>100</u>, <u>121</u>, <u>144</u>.

2. 3, <u>5</u>, <u>7</u>, <u>9</u>, 11, <u>13</u>, 15, <u>17</u>, 19, 21, <u>23</u>.

3. It is arithmetic (the sequence of odd numbers.)

4. 2.

5. 1, 4, 9, 16, 25.

6. The sequence of squares.

7. 1, 3, 5, 7, 9.

8. The sequence of odd numbers.

9. 2 meters per second per second.

10. 4, 8, <u>12</u>, <u>16</u>, <u>20</u>, <u>24</u>.

11. Arithmetic.

12. 1, 4, <u>9</u>, <u>16</u>, <u>25</u>, <u>36</u>.

13. The sequence of squares.

14. 4, 8, 16, 32.

15. It is doubled.

16. 1, 4, 16, 64.

17. It is multiplied by 4.

18. 1, 3, 6, 10, 15, 21, 28, 36, 45, 55.

19.
$$1 = 1$$
$$1 + 2 = \underline{3}$$
$$1 + 2 + 3 = \underline{6}$$
$$1 + 2 + 3 + 4 = \underline{10}$$
$$1 + 2 + 3 + 4 + 5 = \underline{15}$$

20. Triangular numbers.

21. 1, 3, 6, 10, 15, 21, 28, 36, 45, 55;
4, 9, 16, 25, 36, 49, 64, 81, 100.

22. Its terms are all squares.

23. Each square number larger than 1 is the sum of two consecutive triangular numbers.

Set II

1. No.

2. No.

3. 9.

4. 1, 5, 6, and 0.

5. No. 2, 3, 7, and 8.

6. 3,721. The other three numbers end in digits in which square numbers do not end.

7. Yes. Examples are 64, 400, and 484.

8. No.

9.
Number	Square
51	2,601
52	2,704
53	2,809
54	2,916
55	3,025
56	3,136
57	3,249
58	3,364
59	3,481

10. (One pattern: The first two digits of the square numbers are 26, 27, 28, ... , 34. The last two digits are squares: 01, 04, 09, ... , 81.)

11. 1, 4, 9, 6, 5, 6, 9, 4, 1.

12. A palindrome reads the same forward and backward.

13.

Number	Square	Digital root of square
1	1	1
2	4	4
3	9	9
4	16	7
5	25	7
6	36	9
7	49	4
8	64	1
9	81	9
10	100	1

14.

Number	Square	Digital root of square
11	121	4
12	144	9
13	169	7
14	196	7
15	225	9
16	256	4
17	289	1
18	324	9
19	361	1
20	400	4

15. No. 1, 4, 7, and 9.

16. They are the same forward and backward.

Set III

1.

Subgroup	s	p	d	f
No. of elements	2	6	10	14

2. 18.

3.

Row	1	2	3	4	5	6
No. of elements	2	8	8	18	18	32

4.

Row	1	2	3	4	5	6
Half no. of elem.	1	4	4	9	9	16

5. They are all square numbers.

6. 32.

7. 50.

Chapter 2, Lesson 5 (pp. 94-98)

Set I

1. 1, 8, 27, 64, 125, <u>216</u>, <u>343</u>, <u>512</u>, <u>729</u>, <u>1,000</u>.

2. 1, 7, <u>19</u>, <u>37</u>, <u>61</u>.

3. 27.

4. 64.

5. 125.

6. They are cube numbers.

7. 6, 24, <u>54</u>, <u>96</u>, <u>150</u>.

8. 1, 8, <u>27</u>, <u>64</u>, <u>125</u>.

9.

Length of edge	1	2	4	8	16
Surface area	6	24	96	384	1,536
Volume	1	8	64	512	4,096

10. It is multiplied by 4.

11. It is multiplied by 8.

12. 81.

13.

1^4	2^4	3^4	4^4	5^4	6^4
1	16	81	256	625	1,296

14.

1^4	2^4	3^4	4^4	5^4	6^4
1^2	4^2	9^2	16^2	25^2	36^2

15. 128.

16.

1^7	2^7	3^7	4^7	5^7
1	128	2,187	16,384	78,125

17.

Number	Cube	Digital root of cube
1	1	1
2	8	8
3	27	9
4	64	1
5	125	8
6	216	9
7	343	1
8	512	8
9	729	9
10	1,000	1
11	1,331	8
12	1,728	9

18. The digital roots form the repeating pattern 1, 8, 9.

Set II

1.

$$1 = 1$$
$$3 + 5 = \underline{8}$$
$$7 + 9 + 11 = \underline{27}$$
$$13 + 15 + 17 + 19 = \underline{64}$$

2. The odd numbers (or, an arithmetic sequence).

3. The sequence of cubes.

4.

$$21 + 23 + 25 + 27 + 29 = 125$$
$$31 + 33 + 35 + 37 + 39 + 41 = 216$$

5. Yes.

6.

$$1 = 1$$
$$1 + 8 = \underline{9}$$
$$1 + 8 + 27 = \underline{36}$$
$$1 + 8 + 27 + 64 = \underline{100}$$

7. The sequence of cubes.

8. The sequence of squares.

9.

$$1^3 = 1^2$$
$$1^3 + 2^3 = \underline{3}^2$$
$$1^3 + 2^3 + 3^3 = \underline{6}^2$$
$$1^3 + 2^3 + 3^3 + 4^3 = \underline{10}^2$$

10. Ignore the exponents and add.

11. $1^3 + 2^3 + 3^3 + 4^3 + 5^3 = \underline{15}^2$
     $1^3 + 2^3 + 3^3 + 4^3 + 5^3 + 6^3 = \underline{21}^2$

12. Yes.

13. $3^2 + 4^2 = \underline{5}^2$
     $3^3 + 4^3 + 5^3 = \underline{6}^3$

14. $3^4 + 4^4 + 5^4 + 6^4 = 7^4$

15. No. $2{,}258 \neq 2{,}401$.

Set III

1. His strength would become $12^2 = 144$ times his normal strength. His weight would become $12^3 = 1{,}728$ times his normal weight.

2. They would break because his weight would have increased 12 times as much as the strength of his bones.

3. Her strength would become $\left(\dfrac{1}{10}\right)^2 = \dfrac{1}{100}$ times her normal strength. Her weight would become $\left(\dfrac{1}{10}\right)^3 = \dfrac{1}{1{,}000}$ times her normal weight.

4. A person should shrink. As exercise 3 illustrates, the person's strength would not decrease as much as his or her weight.

Chapter 2, Lesson 6 (pp. 101-106)

Set I

1.

t_1	t_2	t_3	t_4	t_5	t_6	t_7	t_8
1	1	2	3	5	8	13	21

t_9	t_{10}	t_{11}	t_{12}	t_{13}	t_{14}	t_{15}
34	55	89	144	233	377	610

2. They are every third term.

3. They are every fifth term.

4. 144.

5. 1 and 8.

6. 5, 8, and 13.

7. It is the Fibonacci sequence.

8. 21, 34, and 55.

9. A male bee.

10. A female bee.

11. The white keys.

12. 8.

13. Yes.

14. (Examples.) Octagon, octet, octogenarian, octopus.

15. 5.

16. Yes.

17. (Examples.) Pentagon, pentathlon.

Set II

1.
$$1 + 1 = 2$$
$$1 + 1 + 2 = 4$$
$$1 + 1 + 2 + 3 = 7$$
$$1 + 1 + 2 + 3 + 5 = 12$$
$$1 + 1 + 2 + 3 + 5 + 8 = 20$$
$$1 + 1 + 2 + 3 + 5 + 8 + 13 = 33$$
$$1 + 1 + 2 + 3 + 5 + 8 + 13 + 21 = 54$$

2. Each sum is 1 less than a term of the Fibonacci sequence.

3. 143. $(144 - 1.)$

4.
$$1^2 + 1^2 = 2$$
$$1^2 + 2^2 = 5$$
$$2^2 + 3^2 = 13$$
$$3^2 + 5^2 = 34$$
$$5^2 + 8^2 = 89$$
$$8^2 + 13^2 = 233$$

5. The sums are every other term of the Fibonacci sequence.

6. $13^2 + 21^2 = 610$

7. $1^2 + 1^2 + 2^2 + 3^2 + 5^2 + 8^2 = 104 = 8 \cdot 13$

8. 4,895. $(t_{10} \cdot t_{11}.)$

9. 144 and 610.

10. The numbers in the right column are every third term of the Fibonacci sequence.

11. Pattern B.

12.

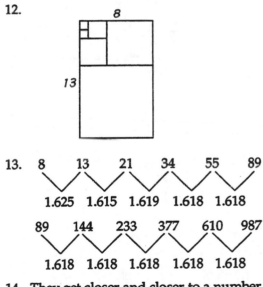

13.

8		13		21		34		55		89
	1.625		1.615		1.619		1.618		1.618	

89		144		233		377		610		987
	1.618		1.618		1.618		1.618		1.618	

14. They get closer and closer to a number around 1.618.

15. A geometric sequence because the ratio between successive terms becomes more and more like a constant ratio.

Set III

1. Five.

2. Other lengths also seem to be approximated by terms of the Fibonacci sequence. There are 10 segments approximately 55 mm long (BG is an example.) There are 15 segments approximately 89 mm long (JF and EF are examples.) There are 5 segments approximately 144 mm long (JG is an example.)

Chapter 2, Review (pp. 108-112)

Set I

1. 0.
2. 1.
3. 2, 6, 10, $\underline{14}$, 18.
4. 9, 16, 25, $\underline{36}$, 49.
5. 4, 12, 36, $\underline{108}$, 324.
6. 1, 8, 27, $\underline{64}$, 125.
7. 8, 13, 21, $\underline{34}$, 55.
8. 1, 3, 6, $\underline{10}$, 15.
9. Arithmetic.
10. Geometric.
11. The number for barley should be 2,401, not 2,301.
12. Binary.
13. Each time the sheets are folded, the number of pages is doubled.
14.

$$1 = 1$$
$$1 + 2 + 1 = \underline{4}$$
$$1 + 2 + 3 + 2 + 1 = \underline{9}$$
$$1 + 2 + 3 + 4 + 3 + 2 + 1 = \underline{16}$$
$$1 + 2 + 3 + 4 + 5 + 4 + 3 + 2 + 1 = \underline{25}$$

15. The sequence of squares.
16. The numbers being added in the pattern are the numbers of dots in the horizontal rows in the figures; the dots in the figures form squares.

Set II

1. $1{,}729 = \underline{1}^3 + \underline{12}^3$.
   $1{,}729 = \underline{9}^3 + \underline{10}^3$.
2. $50 = 1^2 + 7^2$ and $50 = 5^2 + 5^2$.
3. The binary sequence.
4. Red.
5. 7.
6. 5, 17, $\underline{29}$, $\underline{41}$, $\underline{53}$, $\underline{65}$.
7. No. 65 is not prime.
8. 1, $\underline{5}$, 9, $\underline{13}$, $\underline{17}$, $\underline{21}$, $\underline{25}$.
9. 1, $\underline{3}$, 9, $\underline{27}$, $\underline{81}$, $\underline{243}$, $\underline{729}$.
10. 1, $\underline{4}$, 9, $\underline{16}$, $\underline{25}$, $\underline{36}$, $\underline{49}$.
11. The geometric sequence.
12. The arithmetic sequence.
13.

$$5^2 = \underline{25} \qquad 3 \cdot 8 = \underline{24}$$
$$8^2 = \underline{64} \qquad 5 \cdot 13 = \underline{65}$$
$$13^2 = \underline{169} \qquad 8 \cdot 21 = \underline{168}$$
$$21^2 = \underline{441} \qquad 13 \cdot 34 = \underline{442}$$

14. The square of any term seems to differ with the product of the term before it and the term after it by 1. (It is 1 more or 1 less.)
15. $34^2 = 1{,}156 \qquad 21 \cdot 55 = 1{,}155$
16.

$$1^2 + 2^2 + 2^2 = \underline{3}^2$$
$$2^2 + 3^2 + 6^2 = \underline{7}^2$$
$$3^2 + 4^2 + 12^2 = \underline{13}^2$$

17. $4^2 + 5^2 + 20^2 = \underline{21}^2$
18. It is true because $16 + 25 + 400 = 441$.
19. $10^2 + 11^2 + 110^2 = \underline{111}^2$
20. It is true because $100 + 121 + 12{,}100 = 12{,}321$.

Set III

1. A square root of a number is a number that, when squared, gives the number as the result.

2. $74^2 = 5,476$; 5,476 is bigger than 5,248, so 74 is bigger than the square root of 5,248.

3. (Student answer. $72^2 = 5,184$ and $73^2 = 5,329$; 5,184 is closer to 5,248 than 5,329 is, so 72 is a better answer than 73.)

4. (Calculator answers vary. A typical answer: 72.443081.)

5. (Student answer. No, because the square of the number given by the calculator is not 5,248.)

Chapter 2, Further Exploration

Lesson 1 (pp. 113-114)

1. a) 1811, 1828, 1845, 1862, 1879, 1896, 1913, 1930, 1947, 1964, 1981, 1998.
   b) 1801, 1807, 1813, 1819, 1825, 1831, 1837, 1843, 1849, 1855, 1861, 1867, 1873, 1879, 1885, 1891, 1897, 1903, 1909, 1915, 1921, 1927, 1933, 1939, 1945, 1951, 1957, 1963, 1969, 1975, 1981, 1987, 1993, 1999.
   c) 1879 and 1981.
   d) The interval between 1879 and 1981 is 102 years; $102 = 17 \times 6$.
   e) 1805, 1821, 1837, 1853, 1869, 1885, 1901, 1917, 1933, 1949, 1965, 1981, 1997.
   f) 1837, 1885, 1933, and 1981.
   g) 1801, 1819, 1837, 1855, 1873, 1891, 1909, 1927, 1945, 1963, 1981, 1999.
   h) All of them.
   i) The interval in part f is 48 years; $48 = \dfrac{16 \times 6}{2}$. The interval in part h is 18 years; $18 = \dfrac{18 \times 6}{6}$. The interval between the years in which both the locusts and their predators appear seems to be equal to the product of their lifespans divided by their greatest common factor.

2. a) 5,888,896 digits.
   Method:

Numbers	Digits	
1-9	$9 \times 1 =$	9
10-99	$90 \times 2 =$	180
100-999	$900 \times 3 =$	2,700
1,000-9,999	$9,000 \times 4 =$	36,000
10,000-99,999	$90,000 \times 5 =$	450,000
100,000-999,999	$900,000 \times 6 =$	5,400,000
1,000,000		7
		5,888,896

   b) The sum is 27,000,001.
   Method: By pairing the numbers as suggested,

   > 0 and 999,999
   > 1 and 999,998
   > 2 and 999,997 ...

   and looking at the sum of the digits needed to type each pair,

   $$0 + (9 + 9 + 9 + 9 + 9 + 9) = 54$$
   $$1 + (9 + 9 + 9 + 9 + 9 + 8) = 54$$
   $$2 + (9 + 9 + 9 + 9 + 9 + 7) = 54,$$

   we see that the sum of the digits in each pair is 54. The 1,000,000 numbers from 0 through 999,999 form 500,000 such pairs; so the sum of the digits in them is

   $$500,000 \times 54 = 27,000,000.$$

   The sum of the digits needed to type 1,000,000 is 1; so the sum needed to type all of the whole numbers from 1 to 1,000,000 is 27,000,001.

Lesson 2 (pp. 114-115)

1. a) 5,040. $(720 \times 7.)$
   b) 6,227,020,800.

2. a) 0.95.
   b) 81.450625, 77.378093, 73.509188, 69.833728, 66.342041, 63.024938, 59.873691.
   c) 81, 77, 74, 70, 66, 63, 60.
   d) Two.
   e) Seven.

Lesson 3 (pp. 115-117)

1. a) A bushel is 4 times as large as a peck.
   b) 16.
   c) 64.
   d) The smallest unit is the mouthful and the largest is the tun.
   e) 262,144.

2. a) 32. If each card had five holes, the cards could be cut to correspond to the binary numerals from 00000 to 11111, which represent the numbers from 0 to 31.

b) If each card had ten holes, the cards could be cut to correspond to the binary numerals from 0000000000 to 1111111111, which represent the numbers from 0 to 1,023.

Lesson 4 (pp. 117-118)

1. a) It has 9 small squares, 4 "2 × 2" squares and 1 large square.
   b) 30. It has 16 small squares, 9 "3 × 3" squares, 4 "2 × 2" squares, and 1 large square.
   c) 204.

2. The next mileage shown on the odometer that is the same forward and backward would be 16,061. Because

$$16,061 - 15,951 = 110 \text{ and } \frac{110}{2} = 55, \text{ the}$$

average speed of the car in the two hours was 55 miles per hour.

Lesson 5 (pp. 118-119)

1. a) The reporter for the *New York Times* observed that all (positive integer) powers of a number whose last digit is 4 end in either 4 or 6. Also all (positive integer) powers of a number whose last digit is 1 end in 1. Looking at the equation
   $$1,324^n + 731^n = 1,961^n$$
   we see that the three numbers in the equation are either
   $$....4 + ...1 = ...1$$
   or
   $$....6 + ...1 = ...1$$
   But $4 + 1 = 5$, not 1, and $6 + 1 = 7$, not 1; so regardless of the value of n chosen, the equation

$$1,324^n + 731^n = 1,961^n$$
cannot be true.

2. a)

Planet	Distance			Period		
	d	d^2	d^3	p	p^2	p^3
Earth	1	1	1	1	1	1
Mars	1.52	2.3	3.5	1.88	3.5	6.6
Jupiter	5.2	27	140	11.9	140	1,700
Saturn	9.54	91	870	29.5	870	2,600
Uranus	19.2	370	7,100	84	7,100	590,000
Neptune	30.1	910	27,000	165	27,000	4,500,000
Pluto	39.5	1,600	62,000	248	62,000	15,000,000

b) The cube of the distance of a planet from the sun in astronomical units is equal to the square of the planet's period in years. (This is Kepler's Third Law.)

Lesson 6 (pp. 119-120)

1. a) The number of possible paths to a given cell seems to be the Fibonacci number corresponding to the number of the cell. For example, there are 3 paths to the fourth cell and the fourth term of the Fibonacci sequence is 3.
   b) Eight, because the sixth term of the Fibonacci sequence is 8.
   c)

d) 55, because the tenth term of the Fibonacci sequence is 55. (The relation to the Fibonacci sequence can be seen by noting that the number of paths to a cell is simply the sum of the number of paths to the two adjoining cells with lower numbers.)

2. a) (Example).

$$
\begin{array}{cc}
5 & \\
31 & \\
36 & 443 \\
67 & -\ 31 \\
103 & \overline{412} \\
170 & \\
\overline{273} & \\
443 &
\end{array}
$$

b) Direct computation of the terms and the sums of the first few terms as given in the table below shows that the trick works for any a and b, at least for the first 12 terms. This is the sort of general demonstration suitable for most classes.

Numbers	Sums
a	a
b	$a + b$
$a + b$	$2a + 2b$
$a + 2b$	$3a + 4b$
$2a + 3b$	$5a + 7b$
$3a + 5b$	$8a + 12b$
$5a + 8b$	$13a + 20b$
$8a + 13b$	$21a + 33b$
$13a + 21b$	$34a + 54b$
$21a + 34b$	$55a + 88b$
$34a + 55b$	$89a + 143b$
$55a + 89b$	$144a + 232b$

Students with strong backgrounds in algebra may find a formal proof based on the relation $t_n = t_{n+2} - t_{n+1}$.

Adding terms, we get

$$
\begin{aligned}
t_1 &= t_3 - t_2 \\
t_2 &= t_4 - t_3 \\
&\ \vdots \\
t_{n-1} &= t_{n+1} - t_n \\
t_n &= t_{n+2} - t_{n+1} \\
t_1 + t_2 + \cdots + t_n &= t_{n+2} - t_2
\end{aligned}
$$

because the intermediate terms drop out of the sum on the right.

Chapter 3, Lesson 1 (pp. 124-128)

Set I

1. | x | 0 | 1 | 2 | 3 | 4 |
    |---|---|---|---|---|---|
    | y | 4 | 5 | 6 | 7 | 8 |

2. | x | 0 | 1 | 2 | 3 | 4 |
    |---|---|---|---|---|---|
    | y | 0 | 7 | 14 | 21 | 28 |

3. | x | 0 | 1 | 2 | 3 | 4 |
    |---|---|---|---|---|---|
    | y | 8 | 7 | 6 | 5 | 4 |

4. | x | 1 | 2 | 3 | 4 | 5 |
    |---|---|---|---|---|---|
    | y | 12 | 6 | 4 | 3 | 2.4 |

5. | x | 1 | 2 | 3 | 4 | 5 |
    |---|---|---|---|---|---|
    | y | 12 | 23 | 34 | 45 | 56 |

6. | x | 5 | 6 | 7 | 8 | 9 |
    |---|---|---|---|---|---|
    | y | 0 | 2 | 4 | 6 | 8 |

7. | x | 1 | 2 | 3 | 4 | 5 |
    |---|---|---|---|---|---|
    | y | 3 | 3 | 3 | 3 | 3 |

8. | x | 1 | 2 | 3 | 4 | 5 |
    |---|---|---|---|---|---|
    | y | 1 | 4 | 9 | 16 | 25 |

9. | x | 1 | 2 | 3 | 4 | 5 |
    |---|---|---|---|---|---|
    | y | 4 | 9 | 16 | 25 | 36 |

10. | x | 2 | 3 | 4 | 5 | 6 |
    |---|---|---|---|---|---|
    | y | 1 | 8 | 27 | 64 | 125 |

11. | x | 2 | 3 | 4 | 5 | 6 |
    |---|---|---|---|---|---|
    | y | 4 | 8 | 16 | 32 | 64 |

12. Tables 1, (2), 3, 5, 6, and 7.

13. Table 11.

14. Tables 8 and 9.

15. Functions 1, 2, 5, 6, 8, 9, 10, and 11.

16. Functions 3 and 4.

Set II

1. $y = 2x$

2. $y = x + 8$

3. $y = x - 3$

4. $y = x^2$

5. $y = 11x$

6. $y = 10x + 1$

7. $y = 5x - 2$

8. $y = x^3$

9. $y = x^3 + 1$

10. $y = x^2 + x$ or $y = x(x + 1)$

11. 350.

12. By adding 70.

13. By multiplying by 7.

14. $y = 7x$.

15. 660, 840, 1,020, 1,200.

16. Arithmetic.

17. 180.

18. 160.

19. By subtracting 10.

20. By subtracting the corresponding x-number from 220.

21. $y = 220 - x$.

22. | No. of cylinders, x | 4 | 6 | 8 |
    |---|---|---|---|
    | No. of miles per gallon, y | 34 | 29 | 24 |

23. The number of miles per gallon decreases by 5.

Set III

1. Square numbers.

2. Arithmetic.

3. 9 and 2.5.

4. The y-number is half of the square root of the corresponding x-number.

Chapter 3, Lesson 2 (pp. 131-135)

Set I

1. C.
2. B.
3. H.
4. E.
5. K.
6. D.
7. J.
8. A.
9. E.
10. F.

11. A(8, 6), B(4, 5), C(2, 4), D(0, 2), E(-1, 0), F(-2, -4).

12. A(7, 2), B(5, -3), C(3, -6), D(1, -7), E(-1, -6), F(-3, -3), G(-5, 2).

13.

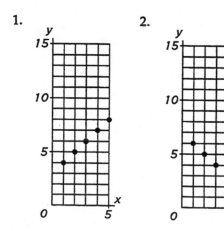

Set II

1.

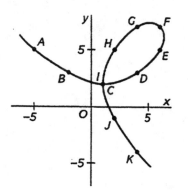

2.

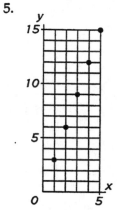

3.

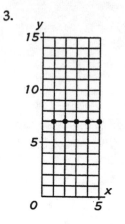

4.

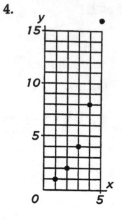

5.

6.

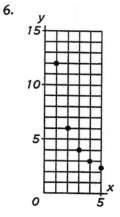

7. A, B, C, and E.

8. A, B, C, and E.

9. D.

10. A, D, and E.

11. B and F.

12. C.

13. B.

14. F.

15. A.

16. E.

Set III

1.

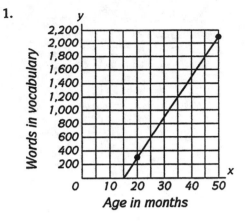

2. 900.

3. 1,500.

4. 60.

5. No. The line meets the *x*-axis at 15. This suggests that a 10-year-old child would know a negative number of words.

Chapter 3, Lesson 3 (pp. 138-144)

Set I

1.
x	0	1	2	3	4
y	3	5	7	9	11

2. Five.

3.

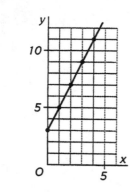

4. $y = x + 4$

5.

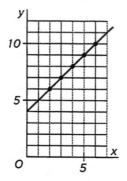

6. At 4.

7. At -2.

8.
x	3	4	5	6	7
y	1	2	3	4	5

9. $y = x - 2$

10. Function A:
x	0	1	2	3	4
y	1	3	5	7	9

Function B:
x	0	1	2	3	4
y	1	4	7	10	13

Function C:
x	0	1	2	3	4
y	1	5	9	13	17

11.

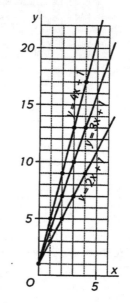

12. (0, 1)

13. 1.

14. $y = 4x + 1$ (Function C).

15. The steepness of the line increases as the number by which x is multiplied in the formula gets larger.

16. Function D:
x	0	1	2	3	4
y	3	4	5	6	7

Function E:
x	0	1	2	3	4
y	5	6	7	8	9

Function F:
x	0	1	2	3	4
y	8	9	10	11	12

17.

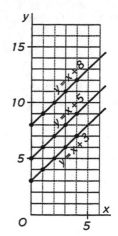

18. The lines are parallel.

19. At 3, 5, and 8.

20. $y = x + 12$.

21. Function G:

x	0	1	2	3	4
y	6	7	8	9	10

Function H:

x	0	1	2	3	4
y	6	5	4	3	2

22.

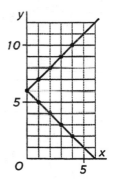

23. A right angle.

Set II

1.

t	0	2	4	6	8	10
h	4	6	8	10	12	14

2. $h = t + 4$

3. 24 centimeters.

4.

t	0	1	2	3	4	5
h	10	8	6	4	2	0

5.

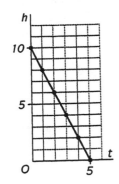

6. 10 centimeters.

7. The height of the candle at the beginning.

8. It decreases by 2 centimeters.

9. The decrease in the height of the candle each hour.

10. Five.

11. It increases the time by 30 minutes.

12.

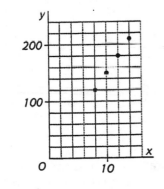

13. $y = 15x$

14.

x	50	60	70	80	90
y	260	320	380	440	500

15. It increases by 60 feet.

16. 560 feet.

17.

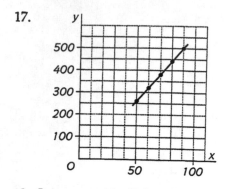

18. It increases the Fahrenheit temperature by 36.

19.

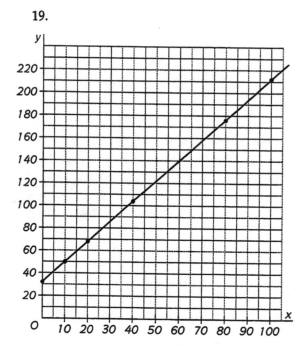

20. 140° F.

21. (One way: In exercise 18, we noticed that increasing the Celsius temperature by 20 increases the Fahrenheit temperature by 36. 40 + 20 = 60 and 104 + 36 = 140.)

Set III

1. $y = 263 - 0.13(1900) = 263 - 247 = 16$

2.

x	1900	1930	1970	1980
y	16	12.1	6.9	5.6

3.

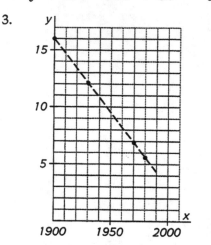

4. Approximately 9.5 feet.

5. No, because the length of a surfboard would never become zero or a negative number.

Chapter 3, Lesson 4 (pp. 146–152)

Set I

1. $y = x^2$.

2.

3.

x	-4	-3	-2	-1
y	16	9	4	1

4. Function A:

x	-3	-2	-1	0	1	2	3
y	11	6	3	2	3	6	11

Function B:

x	-3	-2	-1	0	1	2	3
y	19	14	11	10	11	14	19

5.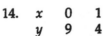

6. They look alike but one is higher than the other.

7. Function A ($y = x^2 + 2$) meets the y-axis at 2; function B ($y = x^2 + 10$) meets the y-axis at 10.

8. $y = x^2 + 7$.

9.

x	-3	-2	-1	0	1	2	3
y	3	8	11	12	11	8	3

10.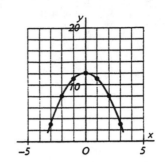

11. It is "upside-down" (or, it opens downward instead of upward.)

12. At 12.

13. $y = 9 - x^2$.

14.

x	0	1	2	3	4	5	6
y	9	4	1	0	1	4	9

15.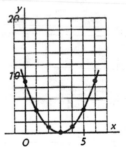

16. At 3.

17. At 9.

18. $y = (5 - x)^2$.

Set II

1.

s	0	1	2	3	4
I	0	2	8	18	32

2. It is multiplied by 4.

3. It is multiplied by 9.

4.

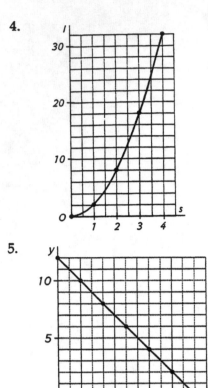

5.

6. The number of subscribers decreases.

7.

x	0	10	20	30	40	50	60
y	0	100	160	180	160	100	0

8.

9. (30, 180)

10. $180,000.

11. $30.

12.

x	0	5	10	15	20	25	27
y	88	85	76	61	40	13	-0.52

13.

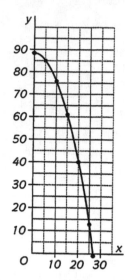

14. The place where the diver hits the water.

15. 88 feet.

16. About 27 feet.

Set III

1.

x	10	12	15	18
y	5.25	6.33	9.00	12.93

2.

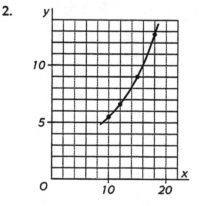

3. $5.37.

4. $7.32.

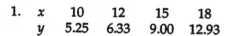

5. Yes (because the curve would move up again on the left.)

6. No! The pizzas in exercises 3 and 4 are smaller than the standard "small" pizza, yet according to the formula, they would have higher prices.

Chapter 3, Lesson 5 (pp. 155-160)

Set I

1.
x	1	2	3	4	5	6
y	6	3	2	1.5	1.2	1

2.

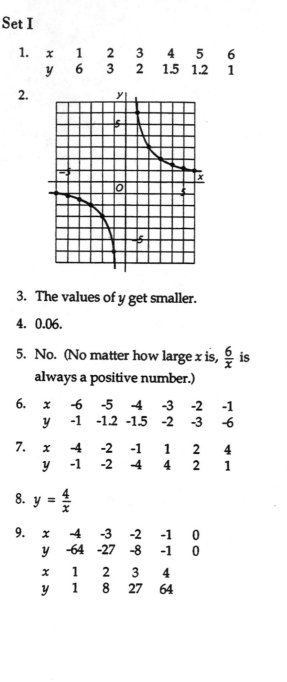

3. The values of y get smaller.

4. 0.06.

5. No. (No matter how large x is, $\frac{6}{x}$ is always a positive number.)

6.
x	-6	-5	-4	-3	-2	-1
y	-1	-1.2	-1.5	-2	-3	-6

7.
x	-4	-2	-1	1	2	4
y	-1	-2	-4	4	2	1

8. $y = \frac{4}{x}$

9.
x	-4	-3	-2	-1	0
y	-64	-27	-8	-1	0

x	1	2	3	4
y	1	8	27	64

10.

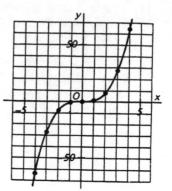

11. The curves have the same shape.

12. One curve is higher than the other.

13. The functions $y = x^6$ and $y = x^8$.

14. The exponents in them are even.

15. The functions $y = x^5$ and $y = x^7$.

16. The exponents in them are odd.

Set II

1. 16.5 meters and 0.0165 meter.

2. It gets shorter.

3.
x	20	40	60	80	100	150
y	16.5	8.25	5.5	4.125	3.3	2.2

4. It is divided in half.

5.

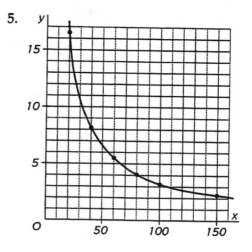

6.

ℓ	1	5	8	10
v	2.2	275	1,126.4	2,200

ℓ	12	15
v	3,801.6	7,425

7. From the table:
   Chicken egg: 275 cubic centimeters,
   Hummingbird egg: 2.2 cubic centimeters.

8. $\dfrac{275}{2.2} = 125$ times the volume.

9. $\dfrac{15}{5} = 3$ times the length.

10. $\dfrac{7,425}{275} = 27$ times the volume.

11.

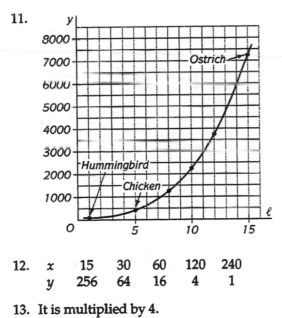

12.

x	15	30	60	120	240
y	256	64	16	4	1

13. It is multiplied by 4.

14.

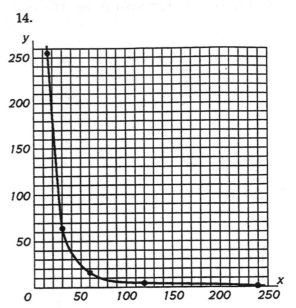

Set III

1. It is multiplied by 4.

2. $y = x^4$.

3.

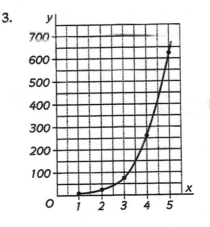

4. $(6)^4 = 1,296$ calories per minute.

Chapter 3, Lesson 6 (pp. 163-167)

Set I

1.

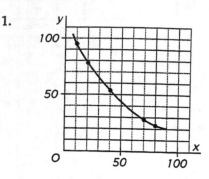

2. A curve.

3. About 40°C.

4. Interpolate. The value estimated is between values that are known.

5. About 20°C.

6. Extrapolate. The value estimated is beyond the values that are known.

7. It slows down.

8.

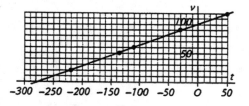

9. They lie along a line.

10. Approximately 80 units.

11. Interpolate.

12. At approximately -273°C.

13. Extrapolate.

Set II

1.

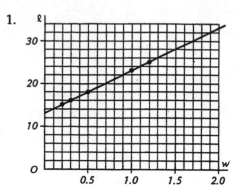

2. 21 centimeters.

3. 13 centimeters.

4. 33 centimeters.

5. The last, because the rubber band might break before that length is reached.

6.

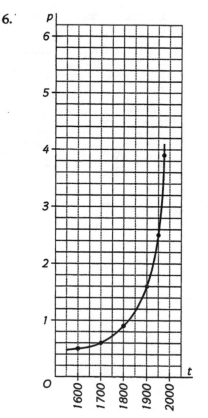

7. It gets steeper and steeper.

8. Approximately 0.4 billion.

9. Approximately 6 billion? (This is hard to estimate.)

10. The estimate for the year 1500.

11. Because the direction of the curve going to the left seems more predictable.

12.

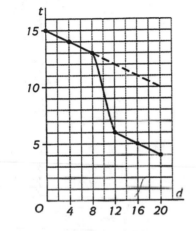

13. 10°C.

14. Between 8 and 12 meters.

15. The lake is only 20 meters deep.

Set III

1. It has increased.

2. 1980.

3.

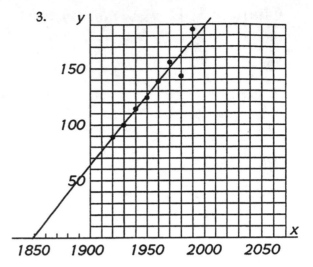

4. For 1970, about 150 mph; for 1980, about 162 mph; for 1990, about 175 mph. (Answers will vary.)

5. About 76 mph. (It was actually 75 mph.)

6. In about 1850.

7. A speed of 0 miles per hour would mean that the winning car isn't even moving! Furthermore, there was no Indianapolis 500 race in 1850 because cars hadn't been invented yet!

8. No. (The winning speeds after 1960 are too unpredictable.)

Chapter 3, Review (pp. 169-173)

Set I

1. (2, 3).

2. (5, -3).

3. (-1, -6).

4. (-4, 0).

5.

x	80	90	100	110
y	50	60	70	80

6. $y = x - 30$.

7. 18.

8. By adding 2.

9. By dividing by 10.

10. $y = \dfrac{x}{10}$.

11. $y = 4x$.

12. $y = x^2$.

13.

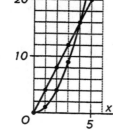

14. Two.

15. (0, 0) and (4, 16).

Set II

1.

x	1	2	3	4	5	6
y	120	60	40	30	24	20

2. It is divided by 2.

3. It is divided by 3.

4.

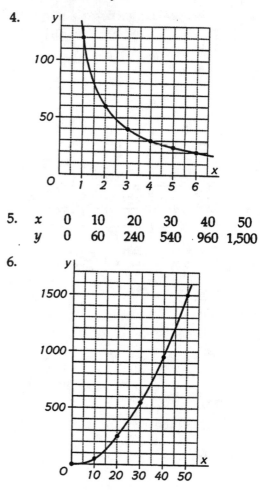

5.

x	0	10	20	30	40	50
y	0	60	240	540	960	1,500

6.

7. About 1,200 feet. (Answers will vary.)

8. 1,215 feet.

9.

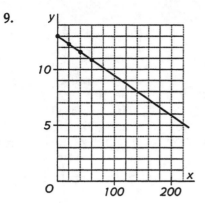

6.

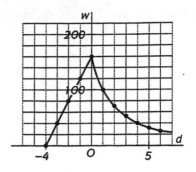

7. At the surface of the earth.

10. About 11.1. (Answers will vary.)

11. Interpolate.

12. About 5.9. (Answers will vary.)

13. Extrapolate.

14. The estimate for the 200th day since September 1 because the days do not keep getting shorter. After the first day of winter, they start getting longer again.

Set III

1.

d	0	1	2	3	4	5	6
w	160	102	71	52	40	32	26

2. It gets smaller.

3. Less than one pound! (By the formula, about 0.88 pound.)

4.

d	0	-1	-2	-3	-4
w	160	120	80	40	0

5. Nothing.

Chapter 3, Further Exploration

Lesson 1 (pp. 174-175)

1. a) To the right.
   b) It increases.
   c) $r = \dfrac{180}{d}$
   d) 72 meters.

   e) No, because the formula requires dividing by the difference and division by 0 is undefined. (In fact, if the difference were exactly 0, the person would not tend to walk in a circle.)

---

2. a) By subtracting the number on the left from the number on the right.

   b)
x	0	1	2	3	4	5	6	7
y	0	1	8	27	64	125	216	343

   1   7   19   37   61   91   127

   6   12   18   24   30   36

   6   6   6   6   6

   c)
x	0	1	2	3	4	5	6	7
y	0	1	16	81	256	625	1,296	2,401

   1   15   65   175   369   671   1,105

   14   50   110   194   302   434

   36   60   84   108   132

   24   24   24   24

   d)
x	0	1	2	3	4	5	6	7
y	0	1	2	3	4	5	6	7

   1   1   1   1   1   1   1

   e) The patterns for $y = x$, $y = x^2$, $y = x^3$, and $y = x^4$ are summarized in the table at the right.

   The corresponding pattern for $y = x^5$ would probably repeat 120's on the 7th row.
   ($120 = 1 \times 2 \times 3 \times 4 \times 5$.)

Function	Patterns	
$y = x$	3rd row repeats 1's	$1 = 1$
$y = x^2$	4th row repeats 2's	$2 = 1 \times 2$
$y = x^3$	5th row repeats 6's	$6 = 1 \times 2 \times 3$
$y = x^4$	6th row repeats 24's	$24 = 1 \times 2 \times 3 \times 4$

   More on the calculus of finite differences, on which this exercise is based, can be found in Martin Gardner's *New Mathematical Diversions from Scientific American* (Simon and Schuster, 1966).

Lesson 2 (pp. 175-177)

1. a)

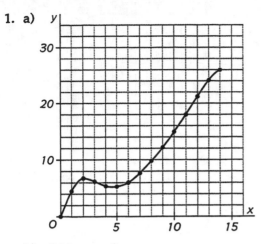

b) 0.14 second.
c) During the first 0.01 second.
d) Between 0.02 and 0.04 second.
e) Between about 0.7 and 0.12 second.
   (Answers will vary.)
f) Slowing down.

2. a)

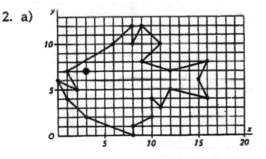

b) (3, 6) (3, 4) (4, 2) (8, 0) (8.5, 1) (11, 2) (12, 4) (13.5, 3) (14.5, 5) (18, 4) (18, 6) (20, 8) (15.5, 7) (13, 8) (16, 10) (15, 12) (13, 10) (14, 12) (11, 10) (4.5, 7) (4.5, 5) (3, 6)

c)

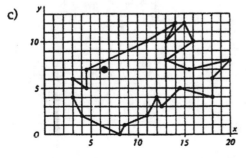

More examples of the applications of coordinates to the study of the shapes of living things can be found in chapter 9 of D'Arcy Thompson's *On Growth and Form*, abridged edition, edited by John Tyler Bonner (Cambridge University Press, 1966).

Lesson 3 (pp. 177-178)

1. a)

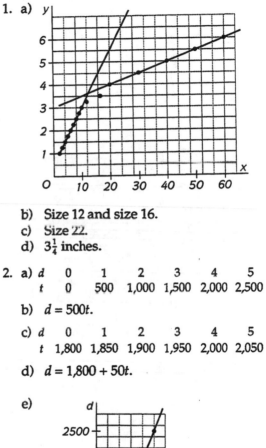

b) Size 12 and size 16.
c) Size 22.
d) $3\frac{1}{4}$ inches.

2. a)

d	0	1	2	3	4	5
t	0	500	1,000	1,500	2,000	2,500

b) $d = 500t$.

c)

d	0	1	2	3	4	5
t	1,800	1,850	1,900	1,950	2,000	2,050

d) $d = 1,800 + 50t$.

e)

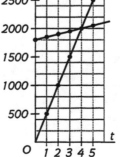

f) Achilles overtakes the tortoise 4 minutes after the start of the race and at 2,000 meters from where he started.

Lesson 4 (pp. 178-179)

1. a)

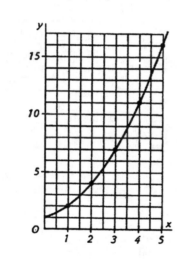

b)

Number of cuts, x	1	2	3	4	5
Number of pieces, y	2	4	7	11	16

c)

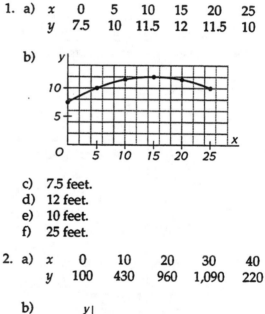

d) If $x = 0$, $y = 1$. Yes, If the number of cuts is 0, the pancake is still in 1 piece.

e) About 8.9. No. There can not be 3.5 cuts.

2. a)

x	-40	-30	-20	-10	0	10	20	30	40
y	52	73	88	97	100	97	88	73	52

b)

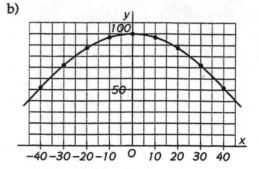

c) From the center.
d) Approximately 48%. (From 100% to 52%.)

Lesson 5 (pp. 179-181)

1. a)

x	0	5	10	15	20	25
y	7.5	10	11.5	12	11.5	10

b)

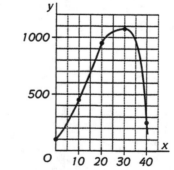

c) 7.5 feet.
d) 12 feet.
e) 10 feet.
f) 25 feet.

2. a)

x	0	10	20	30	40
y	100	430	960	1,090	220

b)

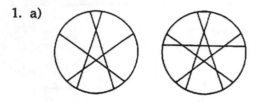

c) First it increases and then it de-
    creases.
d) At 27°C, the brightness of the
    fireflies is approximately 112.9.

"present," and so it is hard to tell
which extrapolation is correct.

Lesson 6 (pp. 181-182)

1. a)

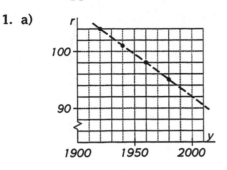

b) The points seem to lie on a line.
c) It seems to decrease at the rate of 3
    males every 20 years.

$$\frac{95}{3} \approx 32; \ 32 \times 20 = 640; \ 1980 + 640$$

    $= 2620$

    If the sex ratio continued to change
    at the same rate, it would become 0
    in about the year 2620. (If this were
    to happen, there would be no males
    left and, hence, eventually no
    population at all.)

2. a) Line a suggests that the universe is
       expanding at a steady rate and that
       it will expand forever.
   b) Curve b suggests that the rate at
       which the universe is expanding is
       slowing down, but that it will
       always continue to expand.
   c) Curve c suggests not only that the
       rate at which the universe is ex-
       panding is slowing down, but also
       that it will eventually collapse.
   d) The first one (because line a meets
       the time axis at the point farthest to
       the left.)
   e) All of them have very nearly the
       same direction at the line marked

Chapter 4, Lesson 1 (pp. 186-191)

Set I

1. 10^0

2. 10^1

3. 10^3

4. 10^{10}

5. 10^{17}

6. 10^{36}

7. One hundred.

8. Ten million.

9. One quintillion.

10. One hundred nonillion.

11. 1,000,000,000 (or one billion).

12. 1.

13. He thought that there were 2 numbers on the board rather than 18.

14. $10^{11} \times 10^{12} = 10^{23}$

15. 10^{25} and 10^{21}

16. 10^{46}

17. 10^7

18. 10^{20}

19. 10^9

20. 10^{34}

21. Add them.

22. 10^{16}

23. 10^{26}

24. 10^{103}

25. Ten thousand, or 10^4, dots.

26. 99,999,999.

27. 10^8

28. 101

29. 10^{100}

30. No.

Set II

1. 10^4

2. One hundred million.

3. 10^8

4. 10,000,000,000,000,000

5. 10^{16}

6. 10^{24}

7. 10^{32}

8. $10^{800,000,000}$

9. 800,000,001

10. No. (Suppose the digits were written nonstop at the rate of one per second. Since 1,000,000 seconds is about 12 days, 800,000,001 seconds is about 800 × 12 = 9,600 days or about 26 years.)

Set III

1.

Seconds	Years	Date*
0	0	[1994]
100 million	3.2	[1991]
1 billion	32	[1962]
10 billion	320	[1674]
100 billion	3,200	[1206 B.C.]
1 trillion	32,000	[30006 B.C.]

*Based on 1994 as example.

2.

Seconds	Years	Date*
0	0	[1994]
100 million	3.2	[1997]
1 billion	32	[2026]
10 billion	320	[2314]
100 billion	3,200	[5194]
1 trillion	32,000	[33994]

*Based on 1994 as example.

Chapter 4, Lesson 2 (pp. 194-198)

Set I

1. 3,000,000,000,000,000,000,000,000,000.
2. 28.
3. 3×10^{27}.
4. 5.
5. Writing it in scientific notation.
6. The coefficient is 2 and the exponent is 15.
7. 2,000,000,000,000,000.
8. Two quadrillion.
9. The coefficient, 635, is not less than 10.
10. 635,000,000,000.
11. 6.35×10^{11}.
12. 330,000,000 and 1,500,000,000.
13. Three hundred thirty million; one billion five hundred million.
14. 1.4×10^5.
15. The biggest number is the one with the biggest exponent. If more than one number have the same exponent, then among them the biggest number is the one with the biggest coefficient.
16. 4×10^8.
17. 200,000,000 and 400,000,000.
18. 2.000744×10^6.
19. 2×10^6.
20. Approximate.

Set II

1. 4,000,000 and 20,000.
2. 80,000,000,000.
3. 8×10^{10}.
4. Multiply the coefficients and add the exponents.
5. 6×10^{15}.
6. 8×10^{20}.
7. $30 \times 10^{11} = 3 \times 10^{12}$.
8. $60 \times 10^{15} = 6 \times 10^{16}$.
9. $81 \times 10^{18} = 8.1 \times 10^{19}$.
10. 5×10^5 and 2.6×10^7.
11. $52 \times 5 \times 10^5 = 260 \times 10^5 = 2.6 \times 10^7$.
12. $15 \times 4 \times 10^{10} = 60 \times 10^{10} = 6 \times 10^{11}$.
13. 15.
14. $\dfrac{7.5 \times 10^5}{5 \times 10^4} = 1.5 \times 10^1$.
15. Divide the coefficients and subtract the exponents.
16. 4×10^2.
17. 1.8×10^{13}.
18. $0.5 \times 10^{35} = 5 \times 10^{34}$.
19. 3×10^{10}.
20. $\dfrac{3 \times 10^{10}}{7.5 \times 10^9} = 0.4 \times 10^1 = 4$.
21. $\dfrac{3 \times 10^{10}}{6.5 \times 10^7} \approx 0.5 \times 10^3 = 500$.
22. 5×10^2, or 500.

23. $\dfrac{500}{60} = 8\dfrac{1}{3}$.

24. $\dfrac{9.3 \times 10^7}{7.7 \times 10^2} \approx 1.2 \times 10^5 = 120{,}000$
hours. (120,000 hours = 5,000 days $\approx$
14 years)

Set III

1. 25¢.

2. $2.34.

3. $930.

4. $250,000,000.

5. $10,000,000,000,000.

Chapter 4, Lesson 3 (pp. 201-206)

Set I

1.
```
      128
   ×   32
      256
      384
     4096
```

2.
```
    128  →     7
     32  →  +  5
  4,096  ←    12
```

3.
```
     1024
   ×  512
     2048
     1024
     5120
   524288
```

4.
```
   1,024  →     10
     512  →  +   9
 524,288  ←     19
```

5.
```
    65,536  →    16
        16  →  +  4
 1,048,576  ←    20
```

6.
```
       8,192  →     13
      16,384  →  +  14
 134,217,728  ←     27
```

7.
```
         64  →     6
        256  →     8
      2,048  →  + 11
 33,554,432  ←    25
```

8. 5.

9. Subtract them.

10.
```
            16
   2048)32768
         2048
        12288
        12288
            0
```

11.
```
  32,768  →     15
   2,048  →  -  11
      16  ←      4
```

12.
```
  262,144  →     18
      256  →  -   8
    1,024  ←     10
```

13.
```
   16,384  →     14
   16,384  →  -  14
        1  ←      0
```

14.
```
  536,870,912  →     29
           64  →  -   6
    8,388,608  ←     23
```

15. 12.

16. Multiply the logarithm of the number by the power.

17. $4^5 = 4 \times 4 \times 4 \times 4 \times 4$
```
       4
   ×   4
      16
   ×   4
      64
   ×   4
     256
   ×   4
    1024
```

18.
```
       4  →     2
          ×     5
   1,024  ←    10
```

19.
```
     128  →     7
          ×     2
  16,384  ←    14
```

20.
$$64 \rightarrow 6$$
$$\underline{\times\ 3}$$
$$262{,}144 \leftarrow 18$$

21.
$$32 \rightarrow 5$$
$$\underline{\times\ 4}$$
$$1{,}048{,}576 \leftarrow 20$$

22.
$$8 \rightarrow 3$$
$$\underline{\times\ 10}$$
$$1{,}073{,}741{,}824 \leftarrow 30$$

Set II

1. 5.

2. 5.

3. 12.

4. 12.

5.

6. $\dfrac{2\times2\times2\times2\times2\times2\times2}{2\times2} = 2\times2\times2\times2\times2.$

7.

8. 5.

9. 5.

10. 12.

11. 12.

12. $2^5 \times 2^7 = 2^{12}.$

13. 5, 7, and 12.

14. They are added.

15. $2^{11} \div 2^3 = 2^8.$

16. 11, 3, and 8.

17. They are subtracted.

18. $(2^4)^3 = 2^{12}.$

19. 4 and 12.

20. It is multiplied by the power.

Set III

1. $2^8 = 256.$

2. $2^{12} = 4{,}096.$

3. $2^{96} \approx 8 \times 10^{28}.$

_____

4. $\dfrac{8 \times 10^{28}}{10^{17}} = 8 \times 10^{11}$ or about 800 billion tons!

Chapter 4, Lesson 4 (pp. 211-215)

Set I

1. $.30 + .78 = 1.08$.

2. 3 and 4.

3. $.48 + .60 = 1.08$.

4. 2 and 7.

5. $.30 + .85 = 1.15$.

6. $.48 + .70 = 1.18$.

7. $.30 + .90 = 1.20$ (or, $.60 + .60 = 1.20$).

8. $.30 + 1 = 1.30$.

9.
Number	10	20	30	40	50
Log	1	1.30	1.48	1.60	1.70
Number	60	70	80	90	100
Log	1.78	1.85	1.90	1.95	2

10.
Number	10	10^2	10^3	10^4	10^5
Log	1	2	3	4	5
Number	10^6	10^7	10^8	10^9	10^{10}
Log	6	7	8	9	10

11. 15.

12. 18.

13. 1.60.

14. 95.

15. One thousand.

16. Ten billion.

17. One nonillion.

Set II

1. $.48 + 6 = 6.48$.

2. $.60 + 8 = 8.60$.

3. $.85 + 5 = 5.85$.

4. $0 + 9 = 9$.

5. $.78 + 12 = 12.78$.

6. 200,000,000.

7. 2×10^8.

8. $.30 + 8 = 8.30$.

9. 8×10^{67}.

10. $.90 + 67 = 67.90$.

11. 1.

12. 10^2 or 100.

13. $8.3 = .30 + 8$; 2×10^8 or 200,000,000.

14. $6.48 = .48 + 6$; 3×10^6 or 3,000,000.

15. $2.85 = .85 + 2$; 7×10^2 or 700.

16. $11.7 = .70 + 11$; 5×10^{11} or 500,000,000,000.

Set III

1. $32 = 4 \times 8$; $.60 + .90 = 1.50$.

2. $\dfrac{1.50}{.60} = 2.5$.

3. $F(5,2) = 2.5$.

4. No. Both decimal and binary logarithms give the same dimension.

Chapter 4, Lesson 5 (pp. 219-223)

Set I

1. .158.

2. .562.

3. .914.

4. 3.158.

5. 8.562.

6. 20.914.

7. 2.2×10^4; 4.342.

8. 3.15×10^{12}; 12.498.

9. 4.5×10^1; 1.653.

10. 1×10^9; 9.000.

11. 1.04.

12. 1.28.

13. 1.48.

14. 1.04×10^5; 104,000.

15. 1.28×10^2; 128.

16. 1.48×10^{10}; 14,800,000,000.

17. 1.05.

18. 1.62.

19. 126.

20. 1,000,000,000,000,000,000,000.

21.

Number	1	2	4	8
Log	0	.301	.602	.903
Number	16	32	64	128
Log	1.204	1.505	1.806	2.107

22. Binary (or geometric).

23. 2.

24. Arithmetic.

25. .301.

26. 2.408 and 2.709.

Set II

1.

Decibels	Loudness	Log of loudness
0	1	0
10	10	1
20	100	2
30	1,000	3
40	10,000	4
50	100,000	5
60	1,000,000	6
70	10,000,000	7
80	100,000,000	8

2. The decibel rating is 10 times the logarithm of its loudness.

3. 100.

4. 1,000,000,000,000.

5. 12.

6. It is 10 times as large.

7. 10^{21} (or, 1,000,000,000,000,000,000,000).

8-9.

Interval of time	Number of seconds	Log
One second	1×10^0	0
One minute	6×10^1	1.8
One hour	3.6×10^3	3.6
One day	8.6×10^4	4.9
One year	3.2×10^7	7.5
One century	3.2×10^9	9.5

10.

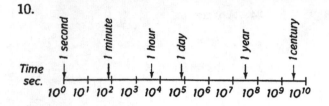

11. It squeezes them together.

12-13.

Wave	Frequency in hertz	Log
AC current	6×10^1	1.8
AM radio station at 980	9.8×10^5	6.0
TV channel 2	5.7×10^7	7.8
Microwave oven	2.45×10^9	9.4
Red light	4×10^{14}	14.6
X-rays	3×10^{16}	16.5

14.

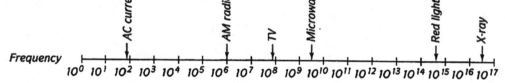

Set III

1. .17.

2. $3.41 + 20(.17) = 3.41 + 3.40 = 6.81$.

3. 6.5×10^6. It had become approximately $6,500,000.

Chapter 4, Lesson 6 (pp. 226-229)

Set I

1. Exponential.

2.
x	0	1	2	3	4	5
y	1	10	100	1,000	10,000	100,000

x	6
y	1,000,000

3.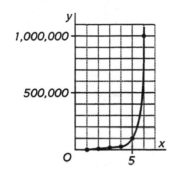

4. It starts out fairly level and then gets steeper and steeper.

5. 25 inches.

6. 250 inches (or nearly 21 feet).

7. The x-numbers are the logarithms of the y-numbers.

8. Exponent.

9.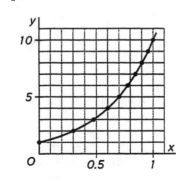

10. Logarithmic.

11. Exponential.

12.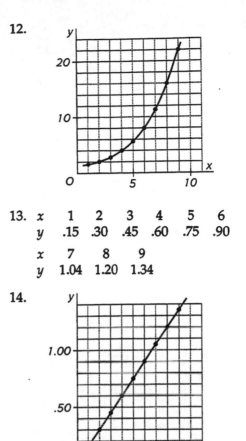

13.
x	1	2	3	4	5	6
y	.15	.30	.45	.60	.75	.90

x	7	8	9
y	1.04	1.20	1.34

14.

15. A line.

Set II

1.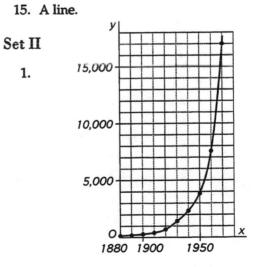

2.

Year	Log	Year	Log
1880	1.5	1930	3.1
1890	1.9	1940	3.3
1900	2.2	1950	3.6
1910	2.5	1960	3.9
1920	2.8	1970	4.2

3.

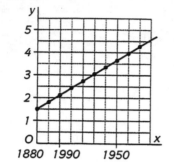

4. They are close to lying on a line.

5.

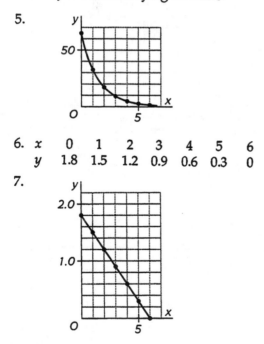

6.

x	0	1	2	3	4	5	6
y	1.8	1.5	1.2	0.9	0.6	0.3	0

7.

8. It changes the graph from a curve to a straight line.

9. About 1.3.

10. $0.2 \times 10^1 = 20$. About 20 grams.

1.

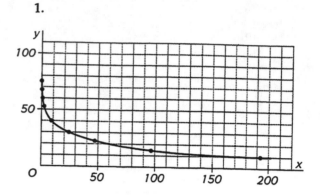

2. It slows down.

3.

Log of time, x	Score, y
0	60
0.3	53
0.9	40
1.4	30
1.7	23
2.0	17
2.3	10

4.

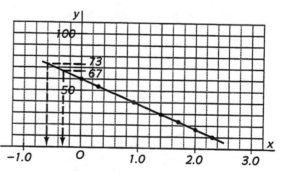

5. They seem to lie on a line.

6. Extending the line to the left, the x-numbers corresponding to 67 and 73 appear to be -0.3 and -0.6. So $\log 0.5 \approx -0.3$ and $\log 73 \approx -0.6$.

Chapter 4, Review (pp. 231-234)

Set I

1. $10,000.

2. About 7,000 pounds.

3. 100,000,000,000,000.

4. One hundred trillion.

5. 1,000,000,000,000,000,000,000,000,000, 000,000,000,000,000,000,000,000,000, 000,000,000,000.

6. 10^{66}.

7. It is shorter to write and easier to read.

8. 4.32×10^9.

9. 1,600,000,000,000.

10. One trillion, six hundred billion.

11. Three.

12.
$$
\begin{array}{r}
729 \\
\times\ 81 \\
\hline
729 \\
5832 \\
\hline
59049
\end{array}
$$

13.
$$
\begin{array}{r}
729 \rightarrow\ 6 \\
81 \rightarrow +\ 4 \\
\hline
59,049 \leftarrow\ 10
\end{array}
$$

14.
$$
\begin{array}{r}
243 \\
9)\overline{2187} \\
\underline{18} \\
38 \\
\underline{36} \\
27 \\
\underline{27} \\
0
\end{array}
$$

15.
$$
\begin{array}{r}
2{,}187 \rightarrow\ 7 \\
9 \rightarrow -\ 2 \\
\hline
243 \leftarrow\ \ \ 5
\end{array}
$$

Set II

1-2.

Creature	Mass in milligrams	Log
Spider	8×10^1	1.9
Hummingbird	2×10^3	3.3
Chicken	3.15×10^6	6.5
Human	7×10^7	7.8
Elephant	6.3×10^9	9.8
Whale	1.38×10^{11}	11.1

3.

4. It squeezes them together.

5. No.

6. Geometric.

7.

x	1	2	3	4	5	6
y	27.5	55	110	220	440	880

x	7	8
y	1,760	3,520

8.

9.

x	1	2	3	4	5	6
y	1.4	1.7	2.0	2.3	2.6	2.9

x	7	8
y	3.2	3.5

10. Arithmetic.

11.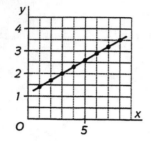

12. They lie on a straight line.

4. $2.3 \rightarrow 2 \times 10^2 = 200.$

5. $\dfrac{200}{100,000} = \dfrac{1}{500}$ or $0.002.$

Set III

1.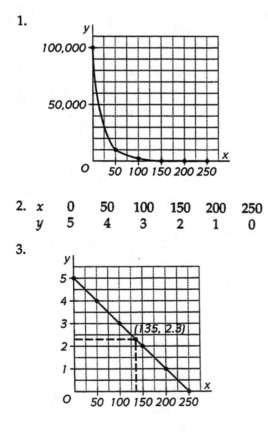

2.

x	0	50	100	150	200	250
y	5	4	3	2	1	0

3.

Chapter 4, Further Exploration

Lesson 1 (pp. 235-236)

1. a) At the word *tryllion*, whose prefix is similar to the Latin word "tria."
   b) 10^{303}. Each power of 10 in Chuquet's system is 3 more than 3 times the number named by the corresponding Latin word:
   $$3 \times 100 + 3 = 303.$$
   c) Perhaps from the facts that our names for large numbers end in "illion" and Z is the last letter of the alphabet.

2. a) 7 seconds.
   b) 10 seconds.
   c) Approximately 11.6 days.
   d) Approximately 11,574 days.
   e) Yes. Only 101 digits are required.
   f) No. A list of all of the numbers from one to a googol would contain 10^{100} numbers. If there are no more than 10^{80} atoms in the universe, even a list made of every atom in the universe would be too small.
   g) No, for the same reason given in part *f*. To print a googolplex in decimal form would require printing $(10^{100} + 1)$ digits.

Lesson 2 (pp. 236-237)

1. a) Geometric.
   b) Arithmetic.
   c) 0.1, 0.01, 0.001, 0.0001. The common ratio of the sequence is 0.1.
   d) 10^{-1}, 10^{-2}, 10^{-3}, 10^{-4}. The common difference of the sequence of exponents is -1.
   e) 5×10^{-9}.
   f) 2×10^{-24} second.

2. a) $9.46 \times 10^{12} \times 10^{7} = 9.46 \times 10^{19}$.
   b) $\dfrac{9.46 \times 10^{19}}{1.70 \times 10^{2}} \approx 5.56 \times 10^{17}$ seconds.
   $$\dfrac{5.56 \times 10^{17}}{3.2 \times 10^{7}} \approx 1.7 \times 10^{10} \text{ years.}$$
   Approximately 17 billion years.

Lesson 3 (pp. 237-239)

1. a) Geometric.
   b)

Number of zone	0	I	II	III	IV
Exposure time	–	–	2^1	2^2	2^3

Number of zone	V	VI	VII	VIII	IX
Exposure time	2^4	2^5	2^6	2^7	2^8

   c) 2^{-1} and 2^{0}.
   d)

Number of zone	0	I	II	III	IV
Log of exp. time	-1	0	1	2	3

Number of zone	V	VI	VII	VIII	IX
Log of exp. time	4	5	6	7	8

   e) The number of each zone is one more than the corresponding logarithm of the exposure time.

2. a) 20 is the logarithm of 1,048,576. 1,048,576 meters $\approx$ 1,000 kilometers $\approx$ 600 miles.
   b) 50 is the logarithm of 1,125,899,906,842,624. 1,125,899,906,842,624 meters $\approx$ 1,000,000,000,000 kilometers $\approx$ 600,000,000,000 miles.
   c) 100 is the logarithm of 1,267,650,600, 228,229,401,496,703,205,376. This is approximately 1,000,000,000,000,000,000,000,000,000 kilometers $\approx$ 600,000,000,000,000,000,000,000,000 miles!

Lesson 4 (p. 239)

1. a) $\log 4 = 0.78$, $\log 6 = 1.00$,
   $\log 8 = 1.17$, $\log 9 = 1.22$,
   $\log 10 = 1.29$.
   b) Six (because the logarithm of 6 is 1; $6 = 6^1$.)
   c) 36.
   d) Three.

2. a)

Age	16	24	36	54	81
Logarithm of age	1.204	1.380	1.556	1.732	1.908

   b) Yes. The common ratio is 1.5. (Multiplying each term in the first line by 1.5 to get the next term is equivalent to adding the logarithm of 1.5, 0.176, to each term in the second line to get the next term.)

Lesson 5 (pp. 240-242)

1. a) To multiply two numbers, such as 2 and 3, on the slide rule, put the 1 on the upper scale directly over the 2 on the lower scale. The answer, 6, appears on the lower scale below the 3 on the upper scale. (The roles of the two scales in doing this can be reversed.)
   (The slide rule multiplies numbers by adding lengths. The lengths are the logarithms of the numbers.)
   b) To divide two numbers, such as 8 by 2, put the 8 on the upper scale directly over the 2 on the lower scale. The answer, 4, appears on the upper scale above the 1 on the lower scale. (Again, the roles of the two scales can be reversed.)
   (The slide rule divides numbers by subtracting lengths.)

2. a) $9^{(9^9)} \approx 9^{385,000,000}$.
   $385,000,000(.954) \approx 367,000,000.$
   367,000,000 is the logarithm of $1 \times 10^{367,000,000}$.
   b) Approximately 367,000,000.

Lesson 6 (pp. 242-243)

1. a)

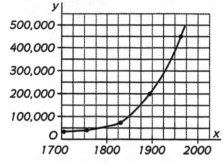

   b) 4.4, 4.6, 4.8, 5.3, 5.7.

   c)

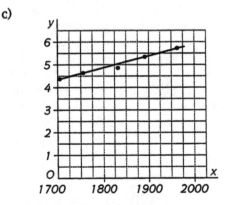

   d) The second graph. (The third point seems to be slightly below the line of the other four points.)
   e) Webster's American Dictionary.

2. a)

Magnitude	6	5	4	3	2	1
Brightness	1	2.5	6.3	16	40	100
Log of brightness	0	0.4	0.8	1.2	1.6	2.0

   b) They form an arithmetic sequence whose common difference is 0.4.

c) 2.4. (Each decrease in the magnitude of one unit corresponds to an
increase in the logarithm of the
brightness of 0.4 unit.)

d) For a magnitude of -12, the logarithm of the brightness would be
$2.4 + 12(0.4) = 7.2$. 7.2 is the logarithm of approximately 1.6×10^7.
The full moon is about 16,000,000
times as bright as a star of the sixth
magnitude.

e) For a magnitude of -27, the logarithm of the brightness would be
$2.4 + 27(0.4) = 13.2$. 13.2 is the
logarithm of approximately
1.6×10^{13}. The sun is about
16,000,000,000,000 times as bright as
a star of the sixth magnitude.

Chapter 5, Lesson 1 (pp. 248-254)

Set I

1. O and Y.

2. E.

3. B.

4. C (or, itself).

5. The point itself.

6. A point on the other side of the line at an equal distance from the line.

7. C.

8. B.

9. E (or, point E doesn't move).

10. The axis of symmetry.

11. O.

12. Water.

13. Yes.

14. No, because it cannot be turned through an angle of less than 360° and look exactly the same.

15. Rotational.

16. So that they look the same when turned upside down.

17. 3.

18. It can be turned through three different angles up to and including 360° so that it looks exactly the same.

19. No.

20. 10.

21. 5.

22. 5-fold.

23. Line symmetry (1 line).

24. Line symmetry (2 lines) and 2-fold rotational symmetry.

25. Line symmetry (2 lines) and 2-fold rotational symmetry.

26. Line symmetry (2 lines) and 2-fold rotational symmetry.

27. No symmetry.

28. Line symmetry (6 lines) and 6-fold rotational symmetry.

29. 8-fold rotational symmetry.

30. Line symmetry (23 lines) and 23-fold rotational symmetry.

Set II

1. Six fish: three white fish facing inward and three brown fish facing outward.

2. A snowflake with 6 lines of symmetry and 6-fold rotational symmetry.

3. A triangle with 3 lines of symmetry and 3-fold rotational symmetry.

4. Same as exercise 3.

5. A star with 5 lines of symmetry and 5-fold rotational symmetry.

6. A square with 4 lines of symmetry and 4-fold rotational symmetry.

7. A "stop sign" (or octagon) with 8 lines of symmetry and 8-fold rotational symmetry.

8. A circle with infinitely many lines of symmetry and infinite rotational symmetry.

9. 3.

10. 360°.

11. 4.

12. 360°.

13.

Figure	A	B	C	D
Number of sides, n	3	4	5	6
Angle of mirrors, m	120°	90°	72°	60°

14. It increases.

15. $m = \dfrac{360}{n}$.

16.

Number of sides, n	8	9	10	12
Angle of mirrors, m	45°	40°	36°	30°

Set III

1. Use one-half of a chessboard instead of an entire one.

2. The board would still look wrong because it would have line symmetry with respect to the mirror. Although a chessboard has rotational symmetry, it does not have line symmetry.

3. Although the little girl is seen reflected in the mirror, the lettering on her sweater is not reversed.

4. One possible explanation: The lettering is reversed on the girl's actual sweater so that it appears normal in a mirror, in the same way that lettering is often reversed on ambulances. The true explanation: The negative of the photograph was turned over when the picture was printed.

Chapter 5, Lesson 2 (pp. 258-265)

Set I

1. 90°.

2.
Number of sides	3	4	5
Mirror angle	120°	90°	72°
Angles of polygon	60°	90°	108°

3. It decreases.

4. It increases.

5. 180°.

6. Regular octagon.

7. $\frac{360°}{8} = 45°$.

8. 135°.

9. Because a regular polygon with 40 sides looks very much like a circle.

10. $\frac{360°}{40} = 9°$.

11. 171°.

12.

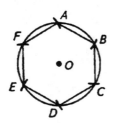

13.

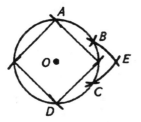

Set II

1. 360.

2. $\frac{360°}{24} = 15°$.

3.

4. Line symmetry (24 lines) and 24-fold rotational symmetry.

5. 15°, 30°, 45°, 60°, 75°, 90°, 105°, etc.

6. $\frac{360°}{3} = 120°$; 120°, 240°, and 360°.

7. Square, hexagon, octagon, and dodecagon.

8. $\frac{360°}{10} = 36°$.

9. 36°, 72°, 108°, 144°, 180°, 216°, 252°, 288°, 324°.

10.

11. A regular pentagon with a five-pointed star inside.

12. Line symmetry (5 lines) and 5-fold rotational symmetry.

13. [Exercise 10 continued.]

14. A regular decagon.

15. $\dfrac{360°}{16} = 22.5°$.

16.

17. Two.

18. One is the wall and the other has its corners at the eight outer pillars.

19. 11.

Set III

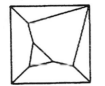

Chapter 5, Lesson 3 (pp. 268-275)

Set I

1. Regular hexagon.

2. Three.

3. | Number of sides, n | 3 | 4 | 5 |
   |---|---|---|---|
   | Mirror angle | 120° | 90° | 72° |
   | Angles of polygon | 60° | 90° | 108° |

6	8	9	10	12
60°	45°	40°	36°	30°
120°	135°	140°	144°	150°

4. Four.

5. $90° + 90° + 90° + 90° = 360°$.

6. Six.

7. $60° + 60° + 60° + 60° + 60° + 60° = 360°$.

8. $108° + 108° + 108° = 324°$;
   $108° + 108° + 108° + 108° = 432°$.

9. Three.

10. 135°.

11. $135° + 135° + 90° = 360°$.

12. An equilateral triangle.

13. $150° + 150° + 60° = 360°$.

14. Three triangles and two squares.

15. Four triangles and one hexagon. Also, two triangles and two hexagons.

16. Not possible.

17. One triangle, two squares, and one hexagon.

18. Not possible.

19. A dodecagon, two triangles, and one square. Also, a dodecagon, a square, and a hexagon.

Set II

1. 6-6-6.

2. 4-8-8.

3. 3-3-3-3-6.

4. 4-6-12.

5. It is not possible because $108° + 120° + 135° = 363°$. The three polygons do not exactly surround the point.

6. It seems possible because $108° + 108° + 144° = 360°$.

7. A, B, D, and E.

8. Too many pentagons meet at point C.

9. No.

10. Yes.

11. No. Too many triangles meet at point C.

12. Yes.

13. 3-6-3-6.

14. 3-4-6-4.

15. Pattern B.

16. Pattern A: 3-3-3-4-4;
    Pattern C: 3-3-4-3-4.

1.

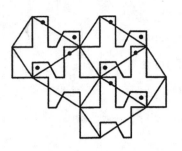

2. The four parts that stick out of the kite
   match the four spaces inside the kite
   that are not filled by the horseman.

3. They are equal.

Set I

1. Cube.

2. Octahedron.

3. Tetrahedron.

4. Icosahedron.

5. Dodecahedron.

6. Four.

7. Four.

8. Six.

9. Three.

10. Two.

11. Six.

12. Eight.

13. Twelve.

14. Three.

15. Two.

16. Eight.

17. Six.

18. Twelve.

19. Squares.

20. Three.

21. Four.

22. Two.

23. Twelve.

24. Twenty.

25. Thirty.

26. Three.

27. Two.

28. Twenty.

29. Twelve.

30. Thirty.

31. Regular pentagons.

32. Twelve.

33. Five.

34. Two.

Set II

1. 3-3-3.

2. 3-3-3-3.

3. 3-3-3-3-3.

4. 3-3-3-3-3-3.

5. 5-5-5.

6. 6-6-6.

7.
Exercise	1	2	3	4	5	6
No. of polygons	3	4	5	6	3	3

8. Three.

9. 180°.

10. 240°.

11. 300°.

12. 324°.

13. It is 360°.

14. It is less than 360°.

15. $6 \times 4 = 24$ corners and $6 \times 4 = 24$ sides;

$\dfrac{24}{3} = 8$ corners and $\dfrac{24}{2} = 12$ edges.

16. $8 \times 3 = 24$ corners and $8 \times 3 = 24$ sides;

$\dfrac{24}{4} = 6$ corners and $\dfrac{24}{2} = 12$ edges.

17. $12 \times 5 = 60$ corners and $12 \times 5 = 60$ sides;

$\dfrac{60}{3} = 20$ corners and $\dfrac{60}{2} = 30$ edges.

18. $20 \times 3 = 60$ corners and $20 \times 3 = 60$ sides;

$\dfrac{60}{5} = 12$ corners and $\dfrac{60}{2} = 30$ edges.

19.

Regular polyhedron	Number of faces	Number of corners	Number of edges
Tetrahedron	4	4	6
Cube	6	8	12
Octahedron	8	6	12
Dodecahedron	12	20	30
Icosahedron	20	12	30

20. The number of edges is two less than the sum of the numbers of faces and corners.

21. An octahedron each of whose six corners touches one of the six faces of a cube.

22. An icosahedron each of whose twelve corners touches one of the twelve faces of a dodecahedron.

23. Yes. One model would consist of a cube each of whose eight corners touches one of the eight faces of an octahedron. Another model would consist of a dodecahedron each of whose twenty corners touches one of the twenty faces of an icosahedron.

Set III

When the tetrahedron is fitted inside the box, each of its six edges lies along one of the six diagonals of the faces of the box.

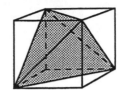

Chapter 5, Lesson 5 (pp. 290-297)

Set I

1. 3-6-6.

2. 3-8-8.

3. 4-6-6.

4. 3-10-10.

5. 5-6-6.

6. 3-4-3-4.

7. 3-4-4-4.

8. 4-6-8.

9. 3-5-3-5.

10. 3-4-5-4.

11. 4-6-10.

12. 3-3-3-3-4.

13. 3-3-3-3-5.

14. Three.

15. An equilateral triangle.

16. A regular octagon.

17. Three.

18. An equilateral triangle.

19. A regular decagon.

20. All of their faces are equilateral triangles.

21. A regular hexagon.

22. Three.

23. Four.

24. Five.

25. Equilateral triangles.

26. Squares.

27. Regular pentagons.

28. The cube and the octahedron.

29. Six.

30. Eight.

31. The icosahedron and the dodecahedron.

32. Twelve.

33. Twenty.

Set II

1. A truncated icosahedron.

2. A great rhombicuboctahedron.

3. Squares, hexagons, and octagons.

4. 12 squares, 8 hexagons, and 6 octagons.

5. 26.

6. 12 separate squares contain
   $12 \times 4 = 48$ corners and sides.
   8 separate hexagons contain
   $8 \times 6 = 48$ corners and sides.
   6 separate octagons contain
   $6 \times 8 = 48$ corners and sides.
   Altogether, this makes 144 separate corners and sides.

   Three corners meet at each corner of the solid, so it has $\frac{144}{3} = 48$ corners.

   Two sides meet at each edge of the solid, so it has $\frac{144}{2} = 72$ edges.

7. 8 separate triangles contain
   $8 \times 3 = 24$ corners and sides.
   6 separate octagons contain
   $6 \times 8 = 48$ corners and sides.
   Altogether, this makes 72 separate
   corners and sides.

   Three corners meet at each corner of
   the solid, so it has $\frac{72}{3} = 24$ corners.

   Two sides meet at each edge of the
   solid, so it has $\frac{72}{2} = 36$ edges.

8. 20 separate triangles contain
   $20 \times 3 = 60$ corners and sides.
   30 separate squares contain
   $30 \times 4 = 120$ corners and sides.
   12 separate pentagons contain
   $12 \times 5 = 60$ corners and sides.
   Altogether, this makes 240 separate
   corners and sides.

   Four corners meet at each corner of
   the solid, so it has $\frac{240}{4} = 60$ corners.

   Two sides meet at each edge of the
   solid, so it has $\frac{240}{2} = 120$ edges.

9. 32 separate triangles contain
   $32 \times 3 = 96$ corners and sides.
   6 separate squares contain
   $6 \times 4 = 24$ corners and sides.
   Altogether, this makes
   120 separate corners and sides.

   Five corners meet at each corner of
   the solid, so it has $\frac{120}{5} = 24$ corners.

   Two sides meet at each edge of the
   solid, so it has $\frac{120}{2} = 60$ edges.

10. The number of edges is two less than
    the sum of the numbers of faces and
    corners.

11. Yes.

12. Yes.

Set III

Chapter 5, Lesson 6 (pp. 300-308)

Set I

1. Equilateral triangles.

2. Rectangles.

3. A triangular prism.

4. Four.

5. Regular tetrahedra.

6. Triangular pyramids.

7. Regular hexagons and rectangles.

8. Squares and rectangles.

9. Hexagonal prisms.

10. Square prisms.

11. Five equilateral triangles.

12. Two equilateral triangles and one regular pentagon.

13. Six.

14. Six.

15. Ten.

16. The figure remains flat rather than three-dimensional.

17. Three. (Triangular, square, and pentagonal pyramids.)

18. A hexagonal pyramid.

19. A square prism.

20. A triangular pyramid.

21. A hexagonal prism.

22. That an equilateral triangle and two squares meet at each corner.

23. 4-4-4, 4-4-5, and 4-4-6.

24. Cube.

25. Yes.

26. A circle.

27. A cone.

28. A cylinder.

Set II

1. 19.

2. 19.

3. 36.

4. Yes, because $19 + 19 = 36 + 2$.

5.

Pyramid	Number of sides in base	Number of faces	Number of corners	Number of edges
Triangular	3	4	4	6
Square	4	5	5	8
Pentagonal	5	6	6	10
Hexagonal	6	7	7	12

6. Yes.

7. $n + 1$.

8. $2n$.

9.

Prism	Number of sides in bases	Number of faces	Number of corners	Number of edges
Triangular	3	5	6	9
Square	4	6	8	12
Pentagonal	5	7	10	15
Hexagonal	6	8	12	18

10. Yes.

11. 12.

12. 20.

13. 30.

14. $n + 2$.

15. $2n$.

16. $3n$.

17. $F + C = E + 2$
$n + 2 + 2n = 3n + 2$
$3n + 2 = 3n + 2$
Yes, it is true.

18. 17.

19. 17.

20. 32.

21. Yes, because $17 + 17 = 32 + 2$.

Set III

1. Pentagonal.

2. A dodecahedron.

3. Twelve.

4. Triangular.

5. An icosahedron.

6. Twenty.

Chapter 5, Review (pp. 310-317)

Set I

1. The heart, spade, and club.

2. The diamond.

3. The diamond.

4. The regular pentagon.

5. 3-12-12.

6. 6-6-6.

7. 4-6-12.

8. 3-3-3-3-6.

9. 360°.

10. Line symmetry.

11. Place a mirror along the vertical line through its center to see if half of the figure is reflected to reproduce the other half. Or, fold the figure along this line to see if the two halves formed coincide.

12. Rotational symmetry.

13. Rotate the figure through an angle of 180° about its center to see if it coincides with its original position.

14. A dodecahedron.

15. Twelve.

16. A regular hexagon.

17. A hexagonal prism.

18. Cube, tetrahedron, dodecahedron. (The remaining two are the icosahedron and octahedron in that order.)

Set II

1. H, I, M, O, T, V, X, and Y. (Also U and W.)

2. C, D, E, H, I, K, O, and X.

3. H, I, N, O, S, X, and Z.

4. A tetrahedron; a triangular pyramid.

5. It has 4 faces, 4 corners, and 6 edges.

6. Yes, because 4 + 4 = 6 + 2.

7.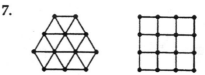

8. Cube, 4-4-4.

9. Truncated cube, 3-8-8.

10. Cuboctahedron, 3-4-3-4.

11. Truncated octahedron, 4-6-6.

12. Octahedron, 3-3-3-3.

13. A dodecagonal prism.

14. Regular dodecagons.

15. Square (or, rectangular.)

16. A hexagonal pyramid.

17. No.

18. Six triangles meet at the top; two triangles and a hexagon meet at each corner of the base.

19. Because of its rotational symmetry, the wrench can be put on it in several different ways.

20. A regular pentagon.

21. Unlike the regular hexagon, the regular pentagon does not have any parallel sides onto which a normal wrench can "grab."

22. The white one is an icosahedron and the brown one is a dodecahedron.

23. Triangular pyramids (brown) and pentagonal pyramids (white).

24. Its symbol, 3-4-4-4, is the same as that of the rhombicuboctahedron. It may have been assumed that only one Archimedean solid could have a given symbol.

25. The numbers of faces are the same, the numbers of edges are the same, and the numbers of corners are the same.

Set III

1. It decreases.

2. No. This pitch is between those of the triangular and square drums. There is no regular polygon whose number of sides is between 3 and 4.

3. Yes. This pitch is between those of the square and hexagonal drums. It might be a pentagonal drum.

4. No. This pitch is less than that of a circular drum. No regular polygon "has more sides" than a circle.

Chapter 5, Further Exploration

Lesson 1 (pp. 318-319)

1. a)

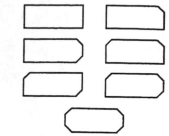

b) Seven.

c) The corners of each bill should be cut so that a person can not deceive someone who is blind by cutting additional corners to make the bill seem more valuable. Consequently, the $1 bill should have all four of its corners cut. The $2 bill should have three corners cut, the $5, $10 and $20 bills should have two corners cut, the $50 bill should have one corner cut, and the $100 bill should have no corners cut.

2. a)

b) (0, -2).

Lesson 2 (pp. 319-320)

1. a) "The Theme" is based on a set of six regular polygons: the equilateral triangle, square, regular pentagon, regular hexagon, regular heptagon, and regular octagon. One side of each polygon is missing and each consecutive pair of polygons shares a side. Starting in the middle, Max Bill began drawing an equilateral triangle, moving clockwise. Instead of drawing the third side, he began a square; instead of drawing the fourth side, he began a regular pentagon; and so forth.

b) "Variation 2" consists of "The Theme" and a set of circles. Each circle has a line segment of "The Theme" as its diameter.

c) In "Variation 5," just one side of each regular polygon is shown, the side that was missing from "The Theme." Also shown are the circles that contain all of each polygon's corners.

2.

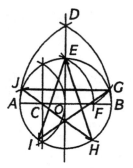

Lesson 3 (pp. 321-322)

1.

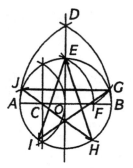

2. a) $\frac{1}{3} + \frac{1}{3} + \frac{1}{3} + \frac{1}{3} + \frac{1}{3} + \frac{1}{3} = 2$

$\frac{1}{3} + \frac{1}{3} + \frac{1}{3} + \frac{1}{3} + \frac{1}{6} = 1\frac{1}{2}$

$\frac{1}{3} + \frac{1}{3} + \frac{1}{3} + \frac{1}{4} + \frac{1}{4} = 1\frac{1}{2}$

$\frac{1}{3} + \frac{1}{3} + \frac{1}{4} + \frac{1}{3} + \frac{1}{4} = 1\frac{1}{2}$

$\frac{1}{3} + \frac{1}{6} + \frac{1}{3} + \frac{1}{6} = 1$

$\frac{1}{3} + \frac{1}{12} + \frac{1}{12} = \frac{1}{2}$

$\frac{1}{4} + \frac{1}{4} + \frac{1}{4} + \frac{1}{4} = 1$

$\frac{1}{4} + \frac{1}{6} + \frac{1}{12} = \frac{1}{2}$

$\frac{1}{4} + \frac{1}{8} + \frac{1}{8} = \frac{1}{2}$

$\frac{1}{6} + \frac{1}{6} + \frac{1}{6} = \frac{1}{2}$

b) The results are summarized in the following table:

Number of polygons	3	4	5	6
Sum of unit fractions	$\frac{1}{2}$	1	$1\frac{1}{2}$	2

An analysis of this pattern is given in *Geometry: An Investigative Approach*, by Phares G. O'Daffer and Stanley R. Clemens (Addison-Wesley, 1976), pages 97–101.

Lesson 4 (pp. 322-323)

1. (It is considerably easier to weave three strips together to make a cube than it is to describe the procedure.)

More information on Jean Pederson's methods for weaving the regular polyhedra can be found in Chapter 10 of *Wheels, Life and Other Mathematical Amusements*, by Martin Gardner (W. H. Freeman, 1983).

---

2. The solution is shown in this figure.

A version of the puzzle having four pieces, each piece consisting of four connected balls, is available in some toy stores.

Lesson 5 (pp. 323-325)

1.

Mosaic	Polyhedron
3-3-3-3-6	3-3-3-3-5
3-4-6-4	3-4-5-4
3-6-3-6	3-5-3-5
4-6-12	4-6-10
4-8-8	3-8-8
5-12-12	5-10-10

2. a)

Polyhedron	Number of corners	Sum of angles around each corner	Angular deficiency
1	12	300°	60°
2	24	330°	30°
3	24	330°	30°
4	60	348°	12°
5	60	348°	12°
6	12	300°	60°
7	24	330°	30°
8	48	345°	15°
9	30	336°	24°
10	60	348°	12°
11	120	354°	6°
12	24	330°	30°
13	60	348°	12°

b) The more corners that a polyhedron has, the smaller its angular defi-

ciency. The formula is $a = \dfrac{720°}{n}$.

2. a) Pentagons and heptagons.
   b) If the cells did not share their edges, there would be $6n$ separate sides. Because two sides come together at each edge of the solid, it must have

   $\dfrac{6n}{2} = 3n$ edges.

   c) Substituting these results into the formula
   $$F + C = E + 2$$
   gives
   $$(n) + (2n) = (3n) + 2.$$
   Because $n + 2n = 3n$, the equation says that a number, $3n$, is 2 more than itself. This is false because *no* number is 2 more than itself.

Lesson 6 (pp 325-326)

1. a)

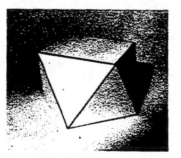

   b) It has two square bases that are parallel.
   c) It has 10 faces, 16 edges, and 8 corners.
   d) It would have $2n + 2$ faces, $4n$ edges, and $2n$ corners.
   e) Yes. $F + C = E + 2$ because
   $(2n + 2) + (2n) = (4n) + 2$.

Chapter 6, Lesson 1 (pp. 330-337)

Set I

PART 1

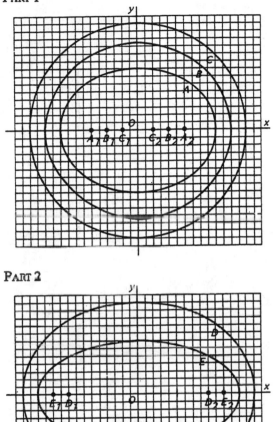

PART 2

PART 3

1. Ellipse C.

2. Ellipse E.

3. Ellipses A and D.

4. Major.

5. 20.

6.

Ellipse	Length of loop used	Distance between foci	Sum of distances of each point on ellipse from foci
A	32	12	20
B	32	8	24
C	32	4	28
D	48	18	30
E	48	22	26

7. They become larger.

8. They become more circular.

9. They become smaller.

10. They become more elongated.

Set II

1. $AF_1 = 3$ cm and $AF_2 = 7$ cm.

2.

Point	Distance from F_1	Distance from F_2
A	3	7
B	5	5
C	9	1
D	8	2
E	4	6

3. 10 cm.

4. The foci.

5. 10 cm.

6. 6 cm.

7. $\dfrac{0^2}{16} + \dfrac{6^2}{36} = 1,$

$\dfrac{0}{16} + \dfrac{36}{36} = 1,$

$0 + 1 = 1.$

8. 36.

9. 6.

10. $\frac{5^2}{25} + \frac{0^2}{25} = 1,$

$\frac{25}{25} + \frac{0}{25} = 1,$

$1 + 0 = 1.$

11. $\frac{(-3)^2}{25} + \frac{4^2}{25} = 1,$

$\frac{9}{25} + \frac{16}{25} = 1,$

$\frac{25}{25} = 1.$

12. $\frac{0^2}{25} + \frac{(-5)^2}{25} = 1,$

$\frac{0}{25} + \frac{25}{25} = 1,$

$0 + 1 = 1.$

13. 25.

14. 5.

15. A circle.

16. An ellipse.

17. Curve B.

18. Curve A: $\frac{x^2}{9} + \frac{y^2}{9} = 1,$

Curve C: $\frac{x^2}{36} + \frac{y^2}{1} = 1.$

Set III

1. Stanton Drew: $\frac{x^2}{169} + \frac{y^2}{25} = 1$

Daviot: $\frac{x^2}{1,369} + \frac{y^2}{144} = 1$

2. Yes. $169 - 25 = 144$ and $144 = 12^2.$

3. $1,369 - 144 = 1,225,\ 1,225 = 35^2.$
The posts were placed at $(-35, 0)$ and $(35, 0).$

Chapter 6, Lesson 2 (pp. 340-346)

Set I

1. Both distances are 6.0 cm.

2. Both distances are 4.9 cm.

3. Both distances are 5.6 cm.

4. The distances from point F are equal to the distances from line ℓ.

5. The focus.

6. The directrix.

7.

x	0	1	2	3	4	5	6
y	0	11	20	27	32	35	36

x	7	8	9	10	11	12
y	35	32	27	20	11	0

8.

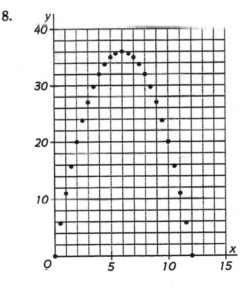

9.

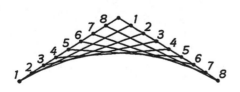

10. A parabola.

11. 20 feet.

12. Approximately 3 feet.

Set II

1.

x	-7	-6	-5	-4	-3	-2	-1	0
y	49	36	25	16	9	4	1	0

x	1	2	3	4	5	6	7
y	1	4	9	16	25	36	49

2.

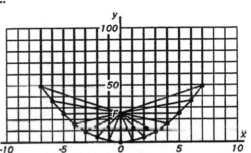

3. They seem to be equal.

4. The farther the ray, the smaller the angle.

5. All of them are reflected in the same direction.

6. The filament at the focus is used for high beam (since it sends the light parallel to the road.)

7. Toward the satellite.

8. A paraboloid.

9. They come together at the focus of the antenna.

Set III

PART 1

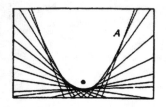

PART 2

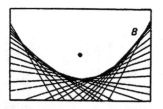

PART 3

1. The foci.

2. It is along the lower edge of the paper.

3. Parabola B.

Chapter 6, Lesson 3 (pp. 350-354)

Set I

PART I

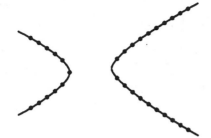

PART 2

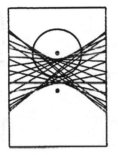

1. One focus is the point that was marked below the circle. The other focus is the center of the circle.

2. Two.

Set II

1. 3 cm.

2.

Point	Distance from F_1	Distance from F_2
A	5.5	3
B	7	4.5
C	2.5	5
D	3.5	6

3. 2.5 cm.

4. 2.5 cm.

5. The foci.

6. The distances get smaller and smaller.

7. $\frac{4^2}{16} - \frac{0^2}{9} = 1$,

$\frac{16}{16} - \frac{0}{9} = 1$,

$1 - 0 = 1$.

8. At 3 and -3.

9. $3^2 = (-3)^2 = 9$; 9 appears below y^2.

10. It crosses the x-axis at -5 and 5 and the y-axis at -2 and 2.

11. $(-5)^2 = 5^2 = 25$; 25 appears below x^2; $(-2)^2 = 2^2 = 4$; 4 appears below y^2.

12. $\frac{x^2}{36} - \frac{y^2}{16} = 1$.

13. $\frac{x^2}{4} - \frac{y^2}{9} = 1$.

Set III

1. Orbit A.

2. Ellipses (orbit C).

Chapter 6, Lesson 4 (pp. 358-364)

Set I

1.

x	0°	15°	30°	45°	60°	75°	90°
$\sin x$	0	0.26	0.50	0.71	0.87	0.97	1.00

2. 135°.

3.

x	105°	120°	135°	150°	165°	180°
$\sin x$	0.97	0.87	0.71	0.50	0.26	0

4. 240°.

5.

x	195°	210°	225°	240°	255°	270°
$\sin x$	-0.26	-0.50	-0.71	-0.87	-0.97	-1.00

6. 315°.

7.

x	285°	300°	315°	330°	345°	360°
$\sin x$	-0.97	-0.87	-0.71	-0.50	-0.26	0

8.

9. Rotational symmetry (2-fold).

10. Around 90° and 270°.

11. Near 0°, 180°, and 360°.

12. Four.

13. 0.

14. -270°, 450°, and 810°.

15. -90°, 630°, and 990°.

Set II

1.

Equation	Amp.	Freq.	Wave
$y = 4 \sin 2x$	4	2	180°
$y = 2 \sin 5x$	2	5	72°
$y = 3 \sin \frac{1}{2}x$	3	$\frac{1}{2}$	720°

2. $y = 4 \sin 2x$.

3. $y = 2 \sin 5x$.

4. $y = 2 \sin 5x$.

5. $y = 3 \sin \frac{1}{2}x$.

6. It decreases.

7. $y = 5 \sin 4x$.

8. $y = \frac{1}{2} \sin 6x$.

9. $y = \sin \frac{3}{4}x$.

10. The amplitude.

11. The frequency (or wavelength).

12. The frequency (or wavelength).

13. The amplitude.

14. Amplitude and frequency. (AM stands for amplitude modulation. FM stands for frequency modulation.)

Set III

1. It decreases.

2. About $\frac{3}{5}$ second.

3. About $\frac{2}{5}$ second.

4. About 100 times.

5. About 150 times.

Chapter 6, Lesson 5 (pp. 367-373)

Set I

1.

Angle	0°	30°	60°	90°	120°	150°	180°
Distance	0	0.5	1	1.5	2	2.5	3

2.

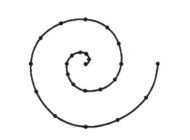

3. 6 feet.

4. Two.

5.

No. of revolutions, x	0	1	2
Distance from center, y	0	6	12

6. Arithmetic.

7. 18 feet and 24 feet.

8.

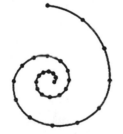

9.

No. of revolutions, x	0	1	2
Distance from center, y	1	3	9

10. Geometric.

11. 27 feet and 81 feet.

Set II

1. 5   10   15   20   25

2. Arithmetic.

3. Archimedean.

4. 50 units.

5. 2   4   8   16

6. Geometric.

7. Logarithmic.

8. 32 units.

9. 3   6   9   12   15   18   21   24

10. Arithmetic.

11. Archimedean.

12. 60 units.

13. 4   16   64

14. Geometric.

15. Logarithmic.

16. 1,024 units.

17. Archimedean.

18. Logarithmic.

Set III

1. The figure seems to be a spiral winding clockwise on a background of other spirals.

2. It actually consists of a set of concentric circles centered on a background of spirals.

Frazer's illusion was first published in the *British Journal of Psychology* in 1908. Good discussions of it and other illusions can be found in *Illusions*, edited by Edi Lanners (Holt, Rinehart and Winston, 1977), and *Seeing: Illusion, Brain, and Mind,* by John P. Frisby (Oxford University Press, 1980).

Chapter 6, Lesson 6 (pp. 376-383)

Set I

PARTS 1, 2, AND 3

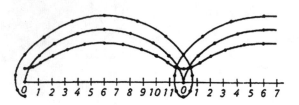

PART 4

1. At the top of the wheel.

2. On the ground.

3. Where it is nearest the ground.

4. No.

5. Yes. (The center of the wheel.)

6. At the bottom of its path the point is going backwards.

7. Below the top of the train track.

Set II

PART 1

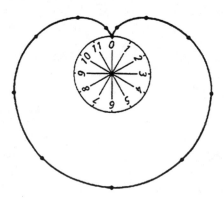

PART 2

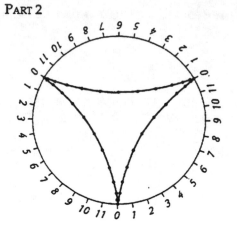

PART 3

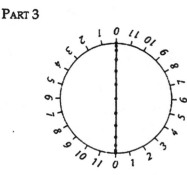

PART 4

1. It is shaped like a heart.

2. One.

3. Line symmetry (one line).

4. No.

5. It has three corners (or, it is "triangular" in shape.)

6. Three.

7. Line symmetry (three lines) and rotational symmetry (threefold).

8. No.

9. A line. (More precisely, a diameter of the circle.)

10. Two.

11. An epicycloid.

12. It is twice as long.

13. Line symmetry (two lines) and rotational symmetry (twofold).

14. A hypocycloid.

15. It is five times as long.

16. Line symmetry (five lines) and rotational symmetry (fivefold).

17. The curve is produced by a point on the rim of a wheel rolling around the inside of another wheel.

18. The curve is produced by a point "outside" a wheel rolling around the outside of another wheel.

Set III

The marble that has to roll farther starts from a point where the curve is steeper and thus it rolls faster.

Chapter 6, Review (pp. 386-392)

Set I

1. The ellipse, circle, parabola, and hyperbola.

2. The conic sections.

3. A sine curve.

4. One wave.

5. A spiral.

6. A cycloid.

7. A hyperbola.

8. A parabola.

9. 75°.

10. 45°.

11. 75°.

12. 15°.

13. A circle.

14. An ellipse.

15. The major axis.

16. At C.

Set II

1. The amplitude of the curve representing the voltage is greater.

2. The wavelengths of the two curves are equal.

3. Six.

4. One-tenth.

5. A spiral.

6. Archimedean.

7. The successive distances of the loops from its center form an arithmetic sequence, (or, its loops are evenly spaced).

8. It is reflected in a single direction.

9. The temperature would rise.

10. Light from the focus is reflected from the mirror in different directions.

11.

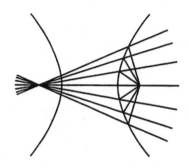

12. The extended lines meet at the other focus of the hyperbola.

13. At the foci.

14.

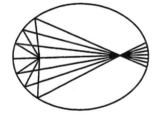

15. Four.

16. Three.

17. Hyperbolas.

18. Four.

19. Point F. (The other points of intersection are on land.)

Set III

1. An ellipse.

2. The minor axis.

3. The direction of the cut.

4. A sine curve.

5. Its wavelength.

6. Its amplitude.

Chapter 6, Further Exploration

Lesson 1 (pp. 393-395)

1. a) An ellipse.

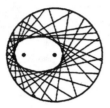

b) At points A and B.
c) It is equal to the radius of the circle.
d) The new fold touches the curve at another point, the sum of whose distances to points A and B is also equal to the radius of the circle.
e) An ellipse is the set of all points in a plane such that the sum of the two distances from each point to two fixed points is the same. This is true of every point on the curve produced in this experiment because the sum of the two distances from each point to points A and B is equal to the radius of the circle.
f) The ellipse becomes rounder as the point is moved closer to the center of the circle, and more elongated as the point is moved closer to the edge.

2. a) Hit the ball so that it goes through the other focus.

b)

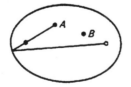

c)

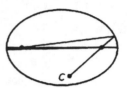

d) It would be back and forth through the foci.

Lesson 2 (pp. 395-396)

a)

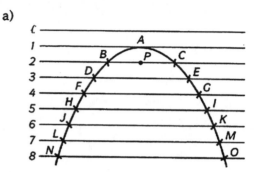

b) At point P.
c) They are the same.
d) A parabola is the set of all points in a plane such that the distance of each point from a fixed point is the same as its distance from a fixed line. The point is P and the line is ℓ.

Lesson 3 (pp. 396-397)

a)

x	$\frac{1}{3}$	$\frac{1}{2}$	1	2	3	4	6	8	10	12
y	13	9	5	3	$2\frac{1}{3}$	2	$1\frac{2}{3}$	$1\frac{1}{2}$	$1\frac{2}{5}$	$1\frac{1}{3}$

b)

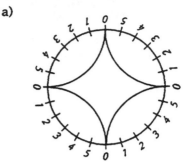

c) To the lower left of the branch graphed (or, in the second and third quadrants.)

d) One asymptote is the *y*-axis. The other asymptote is a line one unit above the *x*-axis.

Lesson 4 (p. 397)

a) 28 days and 33 days.

b) 21,252 days. (As the example in the hint suggests, the number is the least common multiple of the numbers of days in the three cycles. The three numbers 23, 28 and 33 have no common factor other than 1, so their least common multiple is $23 \times 28 \times 33 = 21,252$. This is a little more than 58 years!

Martin Gardner has some interesting things to say about biorhythms in the chapter "The Numerology of Dr. Fliess" in his *Mathematical Carnival* (Knopf, 1975), pages 150-160.

Lesson 5 (p. 398)

a) Logarithmic.
b) 21.
c) 34.
d) The Fibonacci sequence.

Lesson 6 (pp. 399-400)

a)

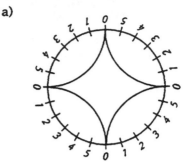

b)

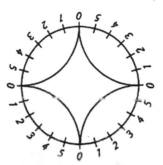

c) The two procedures produce the same curve.

More information on hypocycloids can be found in the chapter on roulettes in *A Book of Curves*, by E. H. Lockwood (Cambridge University Press, 1963).

Chapter 7, Lesson 1 (pp. 404-412)

Set I

1. 3.

2. 3.

3. 18.

4. $2 \times 3 \times 3 = 18$.

5. 9.

6. 6.

7. 9.

8. 3×3.

9. $4 \times 4 = 16$.

10. $5 \times 5 = 25$.

11. $3 \times 3 \times 3 = 27$.

12. $4 \times 4 \times 4 = 64$.

13. $5 \times 5 \times 5 = 125$.

14. 64.

15. $4 \times 4 \times 4 = 64$.

16. $4 \times 4 \times 1 = 16$.

17. $1 \times 4 \times 1 = 4$.

18. $4 \times 4 = 16$.

19. $4 \times 4 \times 4 \times 4 = 256$.

20. $5 \times 5 = 25$.

21. $5 \times 5 \times 5 = 125$.

22. $2 \times 2 \times 2 \times 2 \times 2 \times 2 = 64$.

23. $3 \times 3 \times 3 \times 3 \times 3 \times 3 = 729$.

24. $4 \times 4 \times 4 \times 4 \times 4 \times 4 = 4,096$.

Set II

1. $2 \times 6 \times 20 = 240$.

2. $2 \times 2 \times 1 = 4$.

3. $2 \times 2 \times 2 = 8$.

4. $3 \times 2 \times 1 = 6$.

5. $1 \times 1 \times 1 = 1$.

6. Three sevens.

7. $4 + 8 + 6 + 1 = 19$.

8. $5 \times 5 \times 1 = 25$.

9. $5 \times 5 \times 4 = 100$.

10. $7 \times 3 \times 1 = 21$.

11. $7 \times 3 \times 3 = 63$.

12. $1 \times 5 \times 1 = 5$.

13. $1 \times 5 \times 8 = 40$.

14. $560 + 240 + 19 + 25 + 100 + 21 + 63 + 5 + 40 = 1,073$.

15. $20 \times 20 \times 20 = 8,000$.

16. About once in every eight plays.

17. Plum-cherry-bell.

18. Lemon-lemon-cherry.

19. $6 \times 3 \times 2 = 36$.

20. $36 \times 8 \times 5 = 1,440$.

21. $8 \times 2 \times 9 = 144$.

22. 8.

23. $8 \times 1 \times 9 = 72$.

24. $8 \times 8 \times 10 = 640$.

25. $10 \times 10 \times 10 \times 10 = 10{,}000.$

26. $640 \times 10{,}000 = 6{,}400{,}000.$

27. $2 \times 3 \times 4 = 24.$

28. 24.

29. (One possible answer.) Both problems are about making three choices and the numbers of choices in both problems are the same: 2, 3, and 4. Choosing a child from four children, for example, corresponds to choosing the base of the triangle from the four lines at the bottom of the figure.

Set III

The book contains only 10 right-hand pages. With 10 choices for each line of a poem, there are $10^{14} = 100{,}000{,}000{,}000{,}000$ poems possible.

Chapter 7, Lesson 2 (pp. 416–419)

Set I

1. Permutations.

2. Six gymnasts and three medals.

3. $6 \times 5 \times 4 = 120$.

4. $_7P_5$

5. $7 \times 6 \times 5 \times 4 \times 3 = 2,520$.

6. $_7P_7$

7. $7 \times 6 \times 5 \times 4 \times 3 \times 2 \times 1 = 5,040$.

8. The number of arrangements that can be made from these seven letters taken six at a time.

9. $_9P_2$

10. $9 \times 8 = 72$.

11. $_7P_4$

12. $7 \times 6 \times 5 \times 4 = 840$.

13. $_3P_3$

14. $3 \times 2 \times 1 = 6$.

15. 362,880. ($72 \times 840 \times 6$ or 9!)

16. No. The number 362,880 is too large for the team to be able to play that many games. (Even if they could play 10 games a day, it would take almost 100 years!)

17. 75.

18. 74.

19. 73.

20. $75 \times 74 = 5,550$.

21. $75 \times 74 \times 73 = 405,150$.

Set II

1. False because $6 + 6 \neq 720$.

2. True because $3,628,800 = 10 \times 362,880$.

3. True because $1 + 24 + 120 = 145$.

4. False because $\dfrac{40,320}{24} \neq 2$.

5. True because $720 \times 5,040 = 3,628,800$.

6. $8 \times 7 \times 6 = 336$.

7. 11.

8. $9 \times 8 \times 7 \times 6 \times 5 = 15,120$.

9. $\dfrac{40,320}{120} = 336$.

10. $\dfrac{39,916,800}{3,628,800} = 11$.

11. $\dfrac{362,880}{24} = 15,120$.

12. $\dfrac{15!}{11!}$.

13. $6! = 720$ ways.

14. About 2 years.

15. $13! = 6,227,020,800$.

16. About 197 years! $\left(\dfrac{6,227,020,800}{60 \times 60 \times 24 \times 365} \right)$

17. $5! = 120$.

18. 4, 2, 5, 3, 1.

19. $40 \times 39 \times 38 = 59,280$.

20. $50 \times 49 \times 48 = 117,600$.

Set III

1. $6 \times 5 \times 4 \times 3 \times 2 \times 1 \times 1 = 720$.

2. $1 \times 5 \times 4 \times 3 \times 2 \times 1 \times 1 +$
   $5 \times 4 \times 1 \times 3 \times 2 \times 1 \times 1 +$
   $5 \times 4 \times 3 \times 2 \times 1 \times 1 \times 1 = 360$.

Chapter 7, Lesson 3 (pp. 422-427)

Set I

1. 5,040.

2. 2,520.

3. "Pot-eight-o's."

4. $\dfrac{11!}{9!}$

5. 110.

6. The word ANTEATER has 8 letters, with 2 A's, 2 T's, and 2 E's.

7. 5,040.

8. $\dfrac{8!}{2! \times 2! \times 2!} = \dfrac{8 \times 7!}{2 \times 2 \times 2} = 7!$

9. $\dfrac{13!}{3! \times 3! \times 4! \times 2!}$

10. 3,603,600.

11. $\dfrac{4!}{2! \times 2!}$

12. 6.

13. 1155, 1515, 1551, 5115, 5151, 5511.

14. $\dfrac{5!}{2! \times 3!}$

15. 10.

16.

17. Yes, because there are 10 digits and 10 orders.

18. No, because there are 26 letters but only 10 orders.

Set II

1. 1,024.

2. $\dfrac{10!}{5! \times 5!}$

3. 252.

4. $\dfrac{10!}{6! \times 4!}$

5. 210.

6. $\dfrac{10!}{7! \times 3!}$

7. 120.

8. $\dfrac{10!}{9!}$

9. $\dfrac{10 \times 9!}{9!} = 10.$

10. $\dfrac{10!}{10!} = 1.$

11.

No. of heads	10	9	8	7	6	
No. of tails	0	1	2	3	4	
No. of orders	1	10	45	120	210	

No. of heads	5	4	3	2	1	0
No. of tails	5	6	7	8	9	10
No. of orders	252	210	120	45	10	1

12. (Their sum is 1,024.)

13. $\dfrac{4!}{2! \times 2!} = 6.$

14. $\dfrac{8!}{4! \times 4!} = 70.$

15. $\dfrac{12!}{6! \times 6!} = 924.$

16. Yes, because there are 10 people and only 6 orders.

17. Not very likely, because there are 10 people but 924 orders.

Set III

$$\dfrac{16!}{(2!)^7} = \dfrac{15!}{8} = 163{,}459{,}296{,}000 \text{ ways!}$$

Chapter 7, Lesson 4 (pp. 430-434)

Set I

1. 60.

2. An arrangement of things in a definite order.

3. 60.

4. 10.

5. A selection of things in which the order does not matter.

6. $\frac{60}{6} = 10.$

7. $10 \times 9 = 90.$

8. $\frac{10 \times 9}{2!} = 45.$

9. $12 \times 11 \times 10 = 1,320.$

10. $\frac{12 \times 11 \times 10}{3!} = 220.$

11. $\frac{6}{1!} = 6.$

12. $\frac{6 \times 5}{2!} = 15.$

13. $\frac{6 \times 5 \times 4}{3!} = 20.$

14. $\frac{6 \times 5 \times 4 \times 3}{4!} = 15.$

15. $\frac{6 \times 5 \times 4 \times 3 \times 2}{5!} = 6.$

16. $\frac{6!}{6!} = 1.$

17.

No. of raised dots	1	2	3	4	5	6
No. of ways	6	15	20	15	6	1

18. $6 + 15 + 20 + 15 + 6 + 1 = 63.$

19. $\frac{7 \times 6}{2!} = 21.$

20. $\frac{7 \times 6 \times 5}{3!} = 35.$

21. $\frac{7 \times 6 \times 5 \times 4}{4!} = 35.$

22. $\frac{7 \times 6 \times 5 \times 4 \times 3}{5!} = 21.$

23. $\frac{7 \times 6 \times 5 \times 4 \times 3 \times 2}{6!} = 7.$

24. $\frac{7!}{7!} = 1.$

25. $21 + 35 + 35 + 21 + 7 + 1 = 120.$

Set II

1. $_9C_2 = \frac{9 \times 8}{2!} = 36.$

2. $_{10}C_2 = \frac{10 \times 9}{2!} = 45$, so nine more translators would be needed.

3. $_{16}C_4 = \frac{16 \times 15 \times 14 \times 13}{4!} = 1,820.$

4. $_{12}C_2 = \frac{12 \times 11}{2!} = 66.$

5. $_{24}C_3 = \frac{24 \times 23 \times 22}{3!} = 2,024.$

6. $_{15}C_5 = \frac{15 \times 14 \times 13 \times 12 \times 11}{5!} = 3,003.$

7. $2,024 \times 3,003 = 6,078,072$.

8. $_{36}C_6 = \dfrac{36 \times 35 \times 34 \times 33 \times 32 \times 31}{6!} = $ 1,947,792.

9. $\dfrac{1,947,792}{60 \times 60 \times 24} \approx 22.5$.

   If you bought one ticket per second without stopping, it would take about 22.5 days!

10. $_{16}C_4 = \dfrac{16 \times 15 \times 14 \times 13}{4!} = 1,820$.

11. Exercise 3.

12. The spots on the supermarket cards.

13.

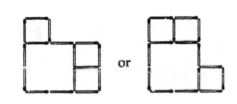

or

14. $_{52}C_5 = \dfrac{52 \times 51 \times 50 \times 49 \times 48}{5!} = $ 2,598,960.

Set III

1. $_{14}C_3 = \dfrac{14 \times 13 \times 12}{3!} = 364$.

2. $_{14}C_4 = \dfrac{14 \times 13 \times 12 \times 11}{4!} = 1,001$.

3. $_{14}C_5 = \dfrac{14 \times 13 \times 12 \times 11 \times 10}{5!} = 2,002$.

4. Inductive.

5. $_{14}C_{13} = \dfrac{14 \times 13 \times 12 \times 11 \times 10 \times 9 \times 8 \times 7 \times 6 \times 5 \times 4 \times 3 \times 2}{13!} = 14$.

6. $_{14}C_6 = 3,003$, $_{14}C_7 = 3,432$, and $_{14}C_8 = 3,003$, so 7 dishes chosen provide the most choices: 3,432.

Chapter 7, Review (pp. 436-440)

Set I

1. $4 \times 4 \times 4 \times 4 \times 4 \times 4 = 4{,}096.$

2. $5 \times 5 \times 5 \times 5 \times 5 \times 5 = 15{,}625.$

3. $5 \times 3 \times 3 = 45.$

4. $7 \times 6 \times 5 = 210.$

5. $2 \times 26 \times 26 = 1{,}352.$

6. $2 \times 26 \times 26 \times 26 = 35{,}152.$

7. No.

8. Yes.

9. $8! = 40{,}320.$

10. $10! = 3{,}628{,}800.$

11. $10 \times 26 \times 10 \times 10 \times 10 \times 10 = 2{,}600{,}000.$

12. $26 \times 26 \times 26 \times 10 \times 10 \times 10 = 17{,}576{,}000.$

13. $10 \times 10 \times 10 \times 26 \times 26 \times 26 = 17{,}576{,}000.$

14. $10 \times 26 \times 26 \times 26 \times 10 \times 10 \times 10 = 175{,}760{,}000.$

15. $14 \times 81 = 1{,}134.$

16. $\dfrac{14 \times 13}{2!} = 91.$

Set II

1. $12 \times 12 \times 12 \times 12 = 20{,}736.$

2. $20 \times 20 \times 20 \times 20 = 160{,}000.$

3. $1 \times 6 \times 1 = 6.$

4. $1 \times 6 \times 5 \times 4 \times 1 = 120.$

5. $1 \times 6 \times 5 \times 4 \times 3 \times 2 \times 1 = 720.$

6. $1 \times 6 \times 5 \times 4 \times 3 \times 2 \times 1 \times 1 = 720.$

7. $\dfrac{7!}{4!} = 210.$

8. $\dfrac{9!}{4! \times 2! \times 2!} = 3{,}780.$

9. $\dfrac{11!}{4! \times 4! \times 2!} = 34{,}650.$

10. $\dfrac{12 \times 11}{2!} = 66.$

11. $\dfrac{12 \times 11 \times 10}{3!} = 220.$

12. 80.

13. $\dfrac{80 \times 79}{2!} = 3{,}160.$

14. $\dfrac{80 \times 79 \times 78}{3!} = 82{,}160.$

15. $\dfrac{80 \times 79 \times 78 \times 77}{4!} = 1{,}581{,}580.$

16.
$$\dfrac{80 \times 79 \times 78 \times 77 \times 76 \times 75 \times 74 \times 73 \times 72 \times 71}{10!}.$$

Set III

1. $4 \times 3 \times 2 \times 1 = 24.$

2. No. (The winners were chosen by a random drawing from correct entries.)

Chapter 7, Further Exploration

Lesson 1 (pp. 441-442)

1. a) $2^{10} = 1,024.$

   b) $\dfrac{1,024}{3} = 341\frac{1}{3}; \; 341 \times \$10 + \$5 = \$3,415.$

   c) (Student response.) [If your solution were the only correct one, you would come out $85 ahead ($3,500 – $3,415). If there was a two-way tie, you would end up losing $1,665 $\left(\dfrac{\$3,500}{2} - \$3,415\right)$.]

2. This table shows the 50 color pairings used in the system.

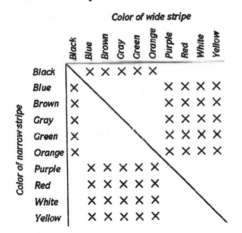

Color of wide stripe

The symmetry of the pairings with respect to the diagonal of the table shows that to each color pairing there corresponds a reversed pairing. In other words, if one wire is blue-white, then another wire is white-blue. One color is chosen from the colors

   blue, brown, gray, green, or orange

and the other color is chosen from the colors

   black, purple, red, white, or yellow.

Lesson 2 (pp. 442-443)

1. a) There are 4! = 24. About 24 seconds.

   b) There are 7! = 5,040. About 84 minutes.

   c) There would be 12! = 479,001,600. No, because at the rate of one change each second, it would take more than 15 years of nonstop ringing.

More information can be found in the article on change ringing in the micropedia of the *Encyclopedia Britannica*.

---

2. a) $\dfrac{4!}{4} = 3! = 6.$

   b) $\dfrac{5!}{5} = 4! = 24.$

   c) $\dfrac{29!}{29} = 28!.$

Lesson 3 (pp. 443-445)

1. a) $\dfrac{5!}{3! \times 2!} = 10.$

   b) $\dfrac{6!}{3! \times 3!} = 20.$

   c)

Ways in which the Yankees win the series

No. of games in the series	4	5	6	7
No. of orders of winners	1	4	10	20

   d) 35.
   e) 35.
   f) 70.

2. a) $\dfrac{5!}{3! \times 2!} = 10.$

b)

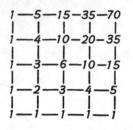

c) Each number along Taylor and O'Farrell is a 1. The rest of the numbers are equal to the sum of the number directly at the left and the number directly below. (The arrangement of the numbers is part of Pascal's triangle, which is the subject of Chapter 8, Lesson 5.)

Lesson 4 (pp. 445-446)

1. a) 210. Each line connects one of the 21 nails to one of the remaining 20. Because the order of the nails is not important for a given line, the number of lines is $_{21}C_2 = \dfrac{21 \times 20}{2!} = 210$.

   b) $_{42}C_2 = \dfrac{42 \times 41}{2!} = 861$.

2. a) $\dfrac{7 \times 6}{2!} = 21$.

   b) $7 + 21 = 28$.

   c) $10 + \dfrac{10 \times 9}{2!} = 55$.

   d) $13 + \dfrac{13 \times 12}{2!} = 91$.

Chapter 8, Lesson 1 (pp. 450-456)

Set I

1. $\frac{1}{6}$.

2. $\frac{1}{6}$.

3. $\frac{3}{6} = \frac{1}{2}$.

4. 0.

5. $\frac{2}{6} = \frac{1}{3}$.

6. $\frac{4}{6} = \frac{2}{3}$.

7. $\frac{1}{5}$.

8. $\frac{4}{5}$.

9. $\frac{1}{3}$.

10. $\frac{2}{3}$.

11. $\frac{1}{7}$.

12. 1.

13. 0.

14. $\frac{6}{7}$.

15. $\frac{2}{8} = \frac{1}{4}$.

16. $\frac{6}{8} = \frac{3}{4}$.

17. $\frac{4}{8} = \frac{1}{2}$.

18. $\frac{1}{8}$.

19. $\frac{7}{8}$.

20. Getting no answer.

21. $\frac{8}{80} = \frac{1}{10}$.

22. 10%.

23. $\frac{10}{80} = \frac{1}{8}$.

24. 12.5%.

25. A game has at least one home run.

26. A game goes into extra innings.

27. Both games of a double-header are won by the same team; the home team wins the game.

28. $\frac{1}{3}$.

29. Twice.

30. 0.

Set II

1. $\frac{1}{2}$, or 0.5.

2. $\frac{5,067}{10,000} = 0.5067$.

3. $\frac{4,933}{10,000} = 0.4933$.

4. That the coin has two heads.

5. 90%.

6. 5%.

7. 4%.

8. No.

9. No, because there are other creatures that can bite.

10. $\frac{1}{38} \approx 2.6\%$.

11. $\frac{18}{38} = \frac{9}{19} \approx 47.4\%$.

12. $\frac{18}{38} = \frac{9}{19} \approx 47.4\%$.

13. $\frac{12}{38} = \frac{6}{19} \approx 31.6\%$.

14. $\frac{2}{38} = \frac{1}{19} \approx 5.3\%$.

15. $\frac{18}{38} = \frac{9}{19} \approx 47.4\%$. (The wheel does not have a memory!)

16. The number 7 because it is the least likely to occur.

Set III

1. $\frac{33}{300} = 0.11$.

2. $\frac{120}{300} = 0.40$.

3. $\frac{180}{300} = 0.60$.

4. 0.

5. $\frac{12}{300} = 0.04$.

6. $\frac{174}{300} = 0.58$.

7. $\frac{126}{300} = 0.42$.

8. 0.

9. 0.

10. $\frac{180}{300} = 0.60$.

11. $\frac{120}{300} = 0.40$.

12. $\frac{30}{300} = 0.10$.

13. The player should aim at A.

14. The player should aim at C.

Chapter 8, Lesson 2 (pp. 459-466)

Set I

1. | Sum of dice | 2 | 3 | 4 | 5 | 6 | 7 |
    |---|---|---|---|---|---|---|
    | No. of ways | 1 | 2 | 3 | 4 | 5 | 6 |

Sum of dice	8	9	10	11	12
No. of ways	5	4	3	2	1

2. $6 + 2 = 8$.

3. $\frac{8}{36} = \frac{2}{9} \approx 22.2\%$.

4. $1 + 2 + 1 = 4$.

5. $\frac{4}{36} = \frac{1}{9} \approx 11.1\%$.

6. $3 + 4 + 5 + 5 + 4 + 3 = 24$.

7. $\frac{24}{36} = \frac{2}{3} \approx 66.7\%$.

8. That a 7 will.

9. 4 and 10.

10. 6 and 8.

11. 49.3%.

12. That the dice thrower will lose.

13.

		First die				
		3	3	4	4	5	5
	1	4	4	5	5	6	6
	1	4	4	5	5	6	6
Second die	5	8	8	9	9	10	10
	5	8	8	9	9	10	10
	6	9	9	10	10	11	11
	6	9	9	10	10	11	11

14. $\frac{4}{36} = \frac{1}{9} \approx 11.1\%$.

15. 0.

16. $\frac{32}{36} = \frac{8}{9} \approx 88.9\%$.

17. 100% because 7 can't turn up.

18. No.

Set II

1. $6 \times 6 \times 6 = 216$.

2. 3.

3. 18.

4. | Sum of dice | 3 | 4 | 5 | 6 | 7 | 8 |
    |---|---|---|---|---|---|---|
    | No. of ways | 1 | 3 | 6 | 10 | 15 | 21 |

Sum of dice	9	10	11	12	13	14
No. of ways	25	27	27	25	21	15

Sum of dice	15	16	17	18
No. of ways	10	6	3	1

5. 3 and 18.

6. 10 and 11.

7. 25 ways to get 9; 27 ways to get 10.

8. 6.

9. 6.

10. 3. (The list is: 1, 4, 4; 4, 1, 4; 4, 4, 1.)

11. 3.

12. 6.

13. 1.

14. $6 + 6 + 3 + 3 + 6 + 1 = 25$.

15. 6.

16. 6.

17. 3.

18. 6.

19. 3.

20. 3.

21. $6+6+3+6+3+3 = 27.$

22.

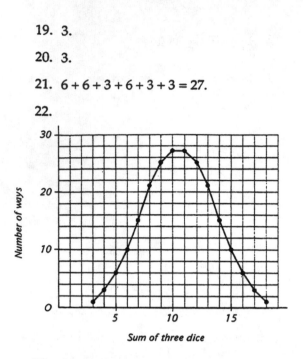

23. $6 \times 6 \times 6 \times 6 \times 6 \times 6 = 46,656.$

24. $\dfrac{1}{46,656}$, or about 0.002%.

25. $50,000.

26. No. (Surprisingly, even if you threw the dice *50,000* times, your probability of winning would be only about $\frac{2}{3}$.)

Set III

1. $\dfrac{11}{18} \approx 61.1\%.$

2.

		Die B				
		0	1	7	8	8	9

Die C	5	C	C	B	B	B	B
	5	C	C	B	B	B	B
	6	C	C	B	B	B	B
	6	C	C	B	B	B	B
	7	C	C	T	B	B	B
	7	C	C	T	B	B	B

3. Player B.

4.

		Die C					
		5	5	6	6	7	7
Die D	3	C	C	C	C	C	C
	4	C	C	C	C	C	C
	4	C	C	C	C	C	C
	5	T	T	C	C	C	C
	11	D	D	D	D	D	D
	12	D	D	D	D	D	D

5. Player C.

6.

		Die D					
		3	4	4	5	11	12
Die A	1	D	D	D	D	D	D
	2	D	D	D	D	D	D
	3	T	D	D	D	D	D
	9	A	A	A	A	D	D
	10	A	A	A	A	D	D
	11	A	A	A	A	T	D

7. Player D.

8. The tables show die A is better than die B, die B is better than die C, die C is better than die D, and die D is better than die A. This means that no die is better than all of the others! No matter which die the first player chooses, the second player can choose one that is more likely to win.

Chapter 8, Lesson 3 (pp. 469-475)

Set I

1. $\frac{1}{2}$.

2. $\frac{1}{2}$.

3. $\frac{1}{2} \times \frac{1}{2} = \frac{1}{4}$.

4. Independent.

5. $\frac{1}{3}$.

6. $\frac{1}{2}$.

7. $\frac{1}{3} \times \frac{1}{2} = \frac{1}{6}$.

8. Dependent.

9. $0.1\% = 0.001 = \frac{1}{1,000}$. One in a thousand.

10. $0.01\% = 0.0001 = \frac{1}{10,000}$. One in ten thousand.

11. $\frac{1}{100} \times \frac{1}{1,000} \times \frac{1}{10,000} = \frac{1}{1,000,000,000}$.

12. $\frac{1}{20} \times \frac{17}{1,000} = \frac{17}{20,000}$.

13. About 17.

14. $\frac{3}{10} \times \frac{1}{2} = \frac{3}{20}$.

15. $\frac{3}{10} \times \frac{1}{2} = \frac{3}{20}$.

16. $\frac{3}{20} \times \frac{3}{20} = \frac{9}{400}$.

17. No. (The lifespans of a husband and wife are influenced by each other.)

18. $\frac{1}{37}$.

19. $\frac{1}{37} \times \frac{1}{37} = \frac{1}{1,369}$.

20. $\frac{1}{37} \times \frac{1}{37} \times \frac{1}{37} = \frac{1}{50,653}$.

21. 0.07.

22. $0.07 \times 0.07 = 0.0049$.

23. $0.07 \times 0.07 \times 0.07 \times 0.07 = 0.00002401$.

24. No. (They might catch colds from each other, they might all go on strike, etc.)

Set II

1. Dependent.

2. $\frac{5}{7}$.

3. In favor of it.

4. $\frac{4}{6} = \frac{2}{3}$.

5. In favor of it.

6. $\frac{3}{5}$.

7. In favor of it.

8. $\frac{5}{7} \times \frac{2}{3} \times \frac{3}{5} = \frac{2}{7}$.

9. Against it.

10. $0.5 \times 0.5 = 0.25$.

11. $0.5 \times 0.5 \times 0.5 = 0.125$.

12. $(0.5)^{12} \approx 0.00024$.

13. $(0.7)^{12} \approx 0.014$.

14. $(0.9)^{12} \approx 0.28$.

15. $0.70 \times 0.75 = 0.525$.

16. $0.60 \times 0.80 = 0.480$.

17. No.

18. Because there are 10 spades in the remaining 47 cards.

19. $\dfrac{9}{46}$.

20. $\dfrac{10}{47} \times \dfrac{9}{46} \approx 4\%$.

Set III

1. 12.

2. $\dfrac{1}{12}$.

3. Dependent.

4. $\dfrac{1}{12} \times \dfrac{1}{12} = \dfrac{1}{144}$.

5. It could make it larger.

6. $\dfrac{1}{12} \times \dfrac{1}{12} \times \dfrac{1}{12} \times \dfrac{1}{12} = \dfrac{1}{20,736}$.

7. No. The wheels of a car do not rotate independently of each other. (On a straight road, all rotate together. In turning a corner, the outside wheels rotate farther than do the inside wheels, but the front and rear wheels on each side rotate about the same amount.)

Chapter 8, Lesson 4 (pp. 478-485)

Set I

1. $0.50 \times 0.50 = 0.25 = 25\%$.

2. $0.50 \times 0.50 = 0.25 = 25\%$.

3.
No. of boys	0	1	2
No. of possibilities	1	2	1
Percent probability	25%	50%	25%

4. $25\% + 25\% = 50\%$.

5. $25\% + 50\% = 75\%$.

6. $0.4 \times 0.4 \times 0.4 = 0.064 = 6.4\%$.

7. $0.4 \times 0.4 \times 0.6 = 0.096 = 9.6\%$.

8. $0.4 \times 0.6 \times 0.6 = 0.144 = 14.4\%$.

9. 3.

10. $3 \times 14.4\% = 43.2\%$.

11. $0.6 \times 0.6 \times 0.6 = 0.216 = 21.6\%$.

12.

No. of hits	0	1	2	3
No. of possibilities	1	3	3	1
Percent probability	21.6%	43.2%	28.8%	6.4%

13. Because there are eight possible outcomes.

14. 100%.

15. 1.

16. 3.

17. $21.6\% + 43.2\% = 64.8\%$.

18. $0.7 \times 0.7 \times 0.7 \times 0.7 = 0.2401 = 24.01\%$.

19. $0.7 \times 0.7 \times 0.7 \times 0.3 = 0.1029 = 10.29\%$.

20. 4.

21. $4 \times 10.29\% = 41.16\%$.

22. $0.7 \times 0.7 \times 0.3 \times 0.3 = 0.0441 = 4.41\%$.

23. 6.

24. $6 \times 4.41\% = 26.46\%$.

25. $0.7 \times 0.3 \times 0.3 \times 0.3 = 0.0189 = 1.89\%$.

26. 4.

27. $4 \times 1.89\% = 7.56\%$.

28. $0.3 \times 0.3 \times 0.3 \times 0.3 = 0.0081 = 0.81\%$.

29.

No. of days that it snows	0	1	2	3	4
No. of possibilities	1	4	6	4	1
Percent probability	1%	8%	26%	41%	24%

30. 16.

31. 100%.

32. 3.

33. $41\% + 24\% = 65\%$.

Set II

1. $\frac{4}{4} = 1 = 100\%$.

2. $\frac{0}{4} = 0 = 0\%$.

3. $\frac{3}{4} = 75\%$.

4. $\frac{1}{4} = 25\%$.

5. $\frac{4}{4} = 1 = 100\%$.

6. $\frac{0}{4} = 0 = 0\%$.

7. $\frac{2}{4} = \frac{1}{2} = 50\%$.

8. Red, $\frac{1}{4} = 25\%$; white, $\frac{1}{4} = 25\%$.

Set III

1.

Number of intersections at which path goes to the left	0	1	2	3	4	5
Number of possibilities	1	5	10	10	5	1
Percent probability	3%	16%	31%	31%	16%	3%

2. A.

3. F.

4. The distances are all the same: five blocks.

5. The beach will be the most crowded in the middle. From the table for exercise 1, we would expect that 3% of the people will be at A, 16% at B, 31% at C, 31% at D, 16% at E, and 3% at F.

Chapter 8, Lesson 5 (pp. 488-494)

Set I

1.

Number of row																					Sum of numbers in row
1									1		1										2
2								1		2		1									4
3							1		3		3		1								8
4						1		4		6		4		1							16
5					1		5		10		10		5		1						32
6				1		6		15		20		15		6		1					64
7			1		7		21		35		35		21		7		1				128
8		1		6		28		56		70		56		28		8		1			256
9	1		9		36		84		126		126		84		36		9		1		512
10	1	10		45		120		210		252		210		120		45		10		1	1,024

2. Binary (or geometric).

3.

No. of teenagers having an accident	0	1	2	3	4	5
No. of ways	1	5	10	10	5	1

4. $0.07776 \approx 7.8\%$.

5. $0.2592 \approx 25.9\%$.

6. 10.

7. $10 \times (0.4 \times 0.4 \times 0.6 \times 0.6 \times 0.6) = 0.3456 \approx 34.6\%$.

8. 10.

9. $10 \times (0.4 \times 0.4 \times 0.4 \times 0.6 \times 0.6) = 0.2304 \approx 23.0\%$.

10. 5.

11. $5 \times (0.4 \times 0.4 \times 0.4 \times 0.4 \times 0.6) = 0.0768 \approx 7.7\%$.

12. $0.4 \times 0.4 \times 0.4 \times 0.4 \times 0.4 = 0.01024 \approx 1.0\%$.

13. Two, because it has the greatest probability.

14.

No. of baskets	0	1	2	3	4	5	6
No. of ways	1	6	15	20	15	6	1

15. $0.9 \times 0.9 \times 0.9 \times 0.9 \times 0.9 \times 0.9 = 0.531441 \approx 53.1\%$.

16. $0.9 \times 0.9 \times 0.9 \times 0.9 \times 0.9 \times 0.1 = 0.059049 \approx 5.9\%$.

17. 6.

18. $6 \times 5.9\% = 35.4\%$.

19. $35.4\% + 53.1\% = 88.5\%$.

20. 20%.

21.

No. of copiers that work	0	1	2	3
No. of ways		1	3	3	1

22. $0.8 \times 0.8 \times 0.8 = 0.512 = 51.2\%$.

23. $0.2 \times 0.2 \times 0.2 = 0.008 = 0.8\%$.

24. $3 \times (0.8 \times 0.8 \times 0.2) = 0.384 = 38.4\%$.

25. $3 \times (0.8 \times 0.2 \times 0.2) = 0.096 = 9.6\%$.

26. $51.2\% + 38.4\% + 9.6\% = 99.2\%$.

27.

No. of correct answers	0	1	2	3	4
No. of ways	1	8	28	56	70

No. of correct answers	5	6	7	8
No. of ways	56	28	8	1

28. $\frac{1}{2} \times \frac{1}{2} \times \frac{1}{2} \times \frac{1}{2} \times \frac{1}{2} \times \frac{1}{2} \times \frac{1}{2} \times \frac{1}{2} = \frac{1}{256}$.

29. 70.

30. $70 \times \frac{1}{256} = \frac{70}{256} \approx 0.273 = 27.3\%$.

31. $28 + 8 + 1 = 37$.

32. $37 \times \frac{1}{256} = \frac{37}{256} \approx 0.145 = 14.5\%$.

Set II

1.

No. of heads	0	1	2	3	4	5
No. of ways	1	10	45	120	210	252
Percent prob.	0%	1%	4%	12%	21%	25%

No. of heads	6	7	8	9	10
No. of ways	210	120	45	10	1
Percent prob.	21%	12%	4%	1%	0%

2.

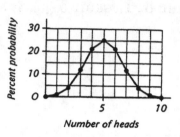

3. 5.

4. It would be better to bet against it, because the probability of its happening is less than 50%.

5. It has a similar shape.

6. It is centered on 10 instead of 5. It does not rise as high in the center.

7. 10.

8. About 18%.

9. When 20 coins are tossed.

10. It gets smaller.

Set III

1. $11^2 = 121$; $11^3 = 1{,}331$; $11^4 = 14{,}641$.

2. They are the second, third, and fourth rows of the triangle.

3. $11^5 = 161{,}051$. No. (The pattern is spoiled because of carrying.)

4. 3, 6, 10, 15, 21.

5. They are triangular numbers.

6. 4, 9, 16, 25, 36.

7. The sequence of squares.

8. 5, 8, 13, 21, 34.

9. The Fibonacci sequence.

Chapter 8, Lesson 6 (pp. 497-503)

Set I

1. $\frac{2}{3}$.

2. Complementary.

3. $\frac{1}{3} \times \frac{1}{3} = \frac{1}{9}$.

4. $1 - \frac{1}{9} = \frac{8}{9}$.

5. $\frac{1}{3} \times \frac{1}{3} \times \frac{1}{3} = \frac{1}{27}$.

6. $1 - \frac{1}{27} = \frac{26}{27}$.

7. $\frac{2}{3} \times \frac{2}{3} \times \frac{2}{3} = \frac{8}{27}$.

8. $1 - \frac{8}{27} = \frac{19}{27}$.

9. 70%.

10. 90%.

11. $0.30 \times 0.10 = 0.03 = 3\%$.

12. $0.70 \times 0.90 = 0.63 = 63\%$.

13. No; one is not the opposite of the other.

14. $\frac{1}{100}$.

15. $\frac{99}{100}$.

16. 97%.

17. $\frac{99}{100} \times \frac{98}{100} \times \frac{97}{100} \approx 0.94 = 94\%$.

18. $\frac{99}{100} \times \frac{98}{100} \times \frac{97}{100} \times \frac{96}{100} \approx 0.90 = 90\%$.

19. Yes.

20. No.

21. 12.

Set II

1. Three out of the ten digits are lucky.

2. $\frac{7}{10}$.

3. $\frac{7}{10} \times \frac{7}{10} \times \frac{7}{10} \times \frac{7}{10} \times \frac{7}{10} \times \frac{7}{10} \times \frac{7}{10} \times \frac{7}{10}$
   $\approx 0.06 = 6\%$.

4. $100\% - 6\% = 94\%$.

5. About 94. (Even more, because most people would have more than one dollar bill.)

6. They realized that almost everyone would have a "lucky dollar" but hoped that people who accepted the prize would pay for more dancing lessons.

7.

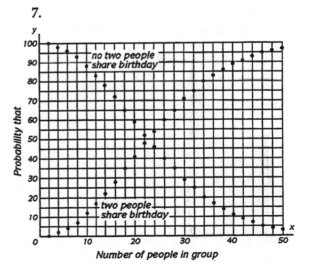

8. At 23 people, the graph shows the probability that two people share the same birthday becoming greater than the probability that no two people do.

9. In favor of it. The table and graph in exercise 7 show that the probability of 2 people in a group of 30 sharing the same birthday is 71%.

10. Yes. Polk and Harding were born on Nov. 2.

11. Telephone numbers can end in 100 different pairs of two digits; the Cracker Jack boxes contain 100 different prizes.

12. Number of telephone numbers.

13. Probability of last two digits being the same.

14. Yes.

15. (Student answer.)

Set III

1. $\frac{5}{6}$.

2. $\frac{5}{6} \times \frac{5}{6} \times \frac{5}{6} \times \frac{5}{6} \approx 0.48 = 48\%$.

3. $100\% - 48\% = 52\%$.

4. In favor of it.

5. $\frac{35}{36}$.

6. $\left(\frac{35}{36}\right)^{24}$.

7. $100\% - 51\% = 49\%$.

8. Against it.

9. No.

Chapter 8, Review (pp. 505-510)

Set I

1. That frogs do not eat dead flies.

2. That you do not get warts from touching a toad.

3. Yes.

4. $100\% - 66\% = 34\%$.

5. $100\% - 84\% = 16\%$.

6. No.

7. $100\% - 21\% = 79\%$.

8. $\dfrac{1}{44}$.

9. $\dfrac{1}{44} \times \dfrac{1}{44} = \dfrac{1}{1,936}$.

10. $\dfrac{1}{44} \times \dfrac{1}{44} \times \dfrac{1}{44} = \dfrac{1}{85,184}$.

11. Pascal's triangle.

12.

```
                    1
                1       1
            1       2       1
        1       3       3       1
    1       4       6       4       1
  1     5      10      10       5      1
 1    6     15     20     15      6      1
1   7    21    35     35     21     7     1
1  8   28   56    70    56    28    8    1
```

13. Binary.

14. $\dfrac{1}{2} \times \dfrac{1}{2} = \dfrac{1}{4}$.

15. $\dfrac{1}{2} \times \dfrac{1}{2} \times \dfrac{1}{2} \times \dfrac{1}{2} \times \dfrac{1}{2} \times \dfrac{1}{2} \times \dfrac{1}{2} \times \dfrac{1}{2} \times \dfrac{1}{2} = \dfrac{1}{512}$.

16. $1 - 0.2 = 0.8$.

17. $0.8 \times 0.8 \times 0.8 \times 0.8 = 0.4096$.

18. $1 - 0.4096 = 0.5904$.

Set II

1. $\dfrac{1}{10}$.

2. $\dfrac{1}{8}$.

3. $\dfrac{1}{10} \times \dfrac{1}{8} = \dfrac{1}{80}$.

4. $\dfrac{1}{9}$.

5. $\dfrac{1}{8}$.

6. $\dfrac{1}{7}$.

7. $\dfrac{1}{9} \times \dfrac{1}{8} \times \dfrac{1}{7} = \dfrac{1}{504}$.

8. $\dfrac{5}{25} = \dfrac{1}{5}$.

9. $\dfrac{1}{5} \times \dfrac{1}{5} \times \dfrac{1}{5} = \dfrac{1}{125}$.

10. No. A probability of $\dfrac{1}{125}$ suggests that 1 person out of 125, or 4 people out of 500, would correctly name three cards in succession solely by luck.

11. $\dfrac{750,000}{25,000,000} = 0.03$.

12. $1 - 0.03 = 0.97$.

13. $0.97 \times 0.97 \times 0.97 \times 0.97 \times 0.97 \times 0.97 \approx$
    0.83.

14. $1 - 0.83 = 0.17$.

15. $6 \times 5 \times 4 = 120$.

16. $\dfrac{120}{6 \times 6 \times 6} = \dfrac{120}{216} = \dfrac{5}{9} \approx 0.56$.

17. You would be more likely to lose.

18. $1 - 0.0001 = 0.9999$.

19. $0.9999 \times 0.9999 \approx 0.9998$.

20. $0.0001 \times 0.0001 = 0.00000001$.

21. $1 - 0.00000001 = 0.99999999$.

Set III

1. $\dfrac{1}{25}$ (or 4%).

2. $\dfrac{1+7}{25+7} = \dfrac{8}{32} = \dfrac{1}{4}$ (or 25%).

3. $\dfrac{1}{32}$ (or about 3%).

4. $\dfrac{8}{32} \times \dfrac{7}{31} = \dfrac{1}{4} \times \dfrac{7}{31} = \dfrac{7}{124}$ (or about
   6%).

5. $\dfrac{8}{31}$ (or about 26%).

Chapter 8, Further Exploration

Lesson 1 (pp. 511-513)

1. a) $\dfrac{2{,}880}{2{,}880+345}=\dfrac{2{,}880}{3{,}225}\approx 89\%.$

   b) $\dfrac{345}{3{,}225}\approx 11\%.$

   c) $\dfrac{2{,}097}{2{,}880}\approx 73\%.$

   d) $\dfrac{289}{345}\approx 84\%.$

   e) Left-handed people.

   f) $\dfrac{2{,}097+289}{3{,}225}\approx 74\%.$

2. a) 4.

   b) $\dfrac{4}{9}.$

   c) $\dfrac{5}{9}.$

   d) Team B.

   e)

Player from		
team B	team C	Match
3	2	3–2
	4	3–4
	9	3–9
5	2	5–2
	4	5–4
	9	5–9
7	2	7–2
	4	7–4
	9	7–9

   f) $\dfrac{4}{9}.$

g) $\dfrac{5}{9}.$

h) Team C.

i) Team C.

j)

Player from		
team A	team C	Match
1	2	1–2
	4	1–4
	9	1–9
6	2	6–2
	4	6–4
	9	6–9
8	2	8–2
	4	8–4
	9	8–9

k) $\dfrac{5}{9}.$

l) $\dfrac{4}{9}.$

m) The answers to parts k and l suggest that team A is better than team C, yet the result of part i implies that team C is the best team! (This strange situation is discussed, together with other nontransitive paradoxes, by Martin Gardner in his book, *Time Travel and Other Mathematical Bewilderments* (W. H. Freeman, 1988).)

Lesson 2 (pp. 513-515)

1. a) $\dfrac{36}{12}=3.$

b)

		First die					
		1	2	3	4	5	6
Second die	0	1	2	3	4	5	6
	0	1	2	3	4	5	6
	0	1	2	3	4	5	6
	6	7	8	9	10	11	12
	6	7	8	9	10	11	12
	6	7	8	9	10	11	12

c) Three faces would have to be blank and three faces would be numbered 6.

2. a and d)

Sum	Times thrown	Winning rolls
2	55	0
3	110	0
4	165	55
5	220	88
6	275	125
7	330	330
8	275	125
9	220	88
10	165	55
11	110	110
12	55	0

b) $\frac{4}{4+6} = \frac{4}{10} = \frac{2}{5}$;

$\frac{2}{5} \times 220 = \frac{440}{5} = 88$.

c) $\frac{5}{5+6} = \frac{5}{11}$; $\frac{5}{11} \times 275 = 125$.

d) [Sum of 8, same calculation as sum of 6.
Sum of 9, same calculation as sum of 5.
Sum of 10, same calculation as sum of 4.]

e) 976.

f) $\frac{976}{1,980} \approx 49.3\%$.

Lesson 3 (pp. 515-517)

1. a) $(0.99)^{51} \approx 0.60 = 60\%$.
   b) $(0.98)^{51} \approx 0.36 = 36\%$.
   c) Yes.
   d) No.
   e) $(0.97)^{51} \approx 0.21 = 21\%$.
   f) $(0.90)^{51} \approx 0.005 = 0.5\%$.

2. a) [Student answer. In the actual experiment, 63% guessed sequence B.]

   b) $\frac{2}{6} = \frac{1}{3}$.

   c) $\frac{4}{6} = \frac{2}{3}$.

   d) $\frac{1}{3} \times \frac{2}{3} \times \frac{1}{3} \times \frac{1}{3} \times \frac{1}{3} = \frac{2}{243} \approx 0.008$.

   e) $\frac{1}{3} \times \frac{2}{3} \times \frac{1}{3} \times \frac{1}{3} \times \frac{1}{3} \times \frac{2}{3} = \frac{4}{729} \approx 0.005$.

   f) $\frac{2}{3} \times \frac{1}{3} \times \frac{1}{3} \times \frac{1}{3} \times \frac{1}{3} \times \frac{1}{3} = \frac{2}{729} \approx 0.003$.

   g) Sequence A, because it has the greatest probability of occurring.

Lesson 4 (pp. 517-521)

1. a) [Student answer. The probability of HTT occurring first 5 or more times out of 10 is 98%!]
   b) [Student answer.]

An analysis of the game, including a table showing the second player's probabilities of winning in each situation, appears in Martin Gardner's *Time Travel* book.

2. a)

Genes of offspring	Color of offspring
BY-BY	Green
BY-BW	Green
BY-WY	Green
BY-WW	Green
BW-BY	Green
BW-BW	Blue
BW-WY	Green
BW-WW	Blue
WY-BY	Green
WY-BW	Green
WY-WY	Yellow
WY-WW	Yellow
WW-BY	Green
WW-BW	Blue
WW-WY	Yellow
WW-WW	White

b) The most common color is green and the least common is white.

c)

Color of parakeet: Green Blue Yellow White

Probability: $\dfrac{9}{16}$ $\dfrac{3}{16}$ $\dfrac{3}{16}$ $\dfrac{1}{16}$

Lesson 5 (pp. 521-523)

1. a) The sixth row.

b) As the first and second numbers.

c) 20.

d) Sixth row, third number.

e) $\dfrac{4 \times 3}{2!} = 6$; fourth row, second number.

f) $\dfrac{7 \times 6 \times 5}{3!} = 35$; seventh row, third number.

g) $\dfrac{8 \times 7 \times 6 \times 5}{4!} = 70$; eighth row, fourth number.

h) $_7C_2$.

i) $\dfrac{7 \times 6}{2!} = 21.$

j) $\dfrac{9 \times 8 \times 7 \times 6 \times 5}{5!} = 126.$

k) On the ninth row as the fifth number.

2. a) The connection is shown in the figure below.

Number of points	Pascal's triangle	Number of regions
2	1 + 1	= 2
3	1 + 2 + 1	= 4
4	1 + 3 + 3 + 1	= 8
5	1 + 4 + 6 + 4 + 1	= 16
6	1 + 5 + 10 + 10 + 5 1	= 31
7	1 + 6 + 15 + 20 + 15 6 1	= 57
8	1 + 7 + 21 + 35 + 35 21 7 1	= 99
9	1 8 28 56 70 56 28 8 1	

b) It would seem to be 163, the sum of the first five numbers of the ninth row.

Lesson 6 (pp. 523-524)

1. a) $0.42 \times 0.5 = 0.21 = 21\%.$

b) $1 - 0.21 = 0.79 = 79\%.$

c) $0.34 \times 0.5 = 0.17 = 17\%.$

d) $1 - 0.17 = 0.83 = 83\%.$

e) 52%.

2. a) $1 - \dfrac{1}{2} \times \dfrac{1}{2} = 1 - \dfrac{1}{4} = \dfrac{3}{4} = 75\%.$

b) $1 - \dfrac{1}{2} \times \dfrac{1}{2} \times \dfrac{1}{2} = 1 - \dfrac{1}{8} = \dfrac{7}{8} = 87.5\%.$

c) No. As long as a power of $\dfrac{1}{2}$ is being subtracted from 1, the result is always less than 1, regardless of how big the power is.

Chapter 9, Lesson 1 (pp. 529-535)

Set I

1. A histogram.

2. 5.

3. 8.

4. No.

5. It has increased.

6.

Distance in meters	Tally marks	Frequency
6.01-6.50	I	1
6.51-7.00		0
7.01-7.50	ⅢⅠ	5
7.51-8.00	IIII	4
8.01-8.50	I	1
8.51-9.00		0

7.

Distance in meters	Tally marks	Frequency
6.01-6.50		0
6.51-7.00		0
7.01-7.50		0
7.51-8.00	II	2
8.01-8.50	IIII	4
8.51-9.00	ⅢⅠ	5

8.

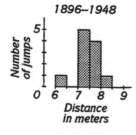

1896–1948

9.

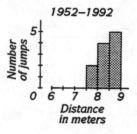

1952–1992

10. It reveals that most of the record distances from 1952 to 1992 were between 8 and 9 meters.

11. About 38%. (Answers will vary.)

12. For an adult to sleep less than 8 hours each night.

13. They are about the same.

14. The writings of Bacon.

15. Shakespeare's plays.

16. Bacon used more words more than 5 letters long than Shakespeare did.

Set II

1.

Age in years	Tally marks	Frequency
20-29		0
30-39	ⅢⅠ ⅢⅠ II	12
40-49	ⅢⅠ ⅢⅠ II	12
50-59	ⅢⅠ	5
60-69	I	1
70-79		0
80-89		0

2.

Age in years	Tally marks	Frequency
20-29	NJ NJ NJ I	16
30-39	NJ NJ	10
40-49	III	3
50-59		0
60-69	I	1
70-79		0
80-89		0

3.

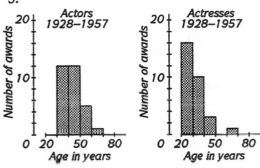

4.

Age in years	Tally marks	Frequency
20-29		0
30-39	NJ III	8
40-49	NJ NJ IIII	14
50-59	IIII	4
60-69	III	3
70-79	I	1
80-89		0

5.

Age in years	Tally marks	Frequency
20-29	NJ I	6
30-39	NJ NJ NJ I	16
40-49	IIII	4
50-59	II	2
60-69	I	1
70-79	I	1
80-89		0

6.

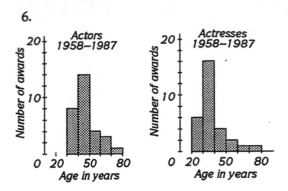

7. The actresses winning the awards are younger than the actors. The winners during the second 30 years of the Awards were older than during the first 30 years.

8. 0 to 1 kilometer.

9. 0 to -1 kilometer and -4 to -5 kilometers.

10. About 30%. (Answers will vary.)

11. About 70%. (Answers will vary.)

12. (The letters that are not in the Hawaiian alphabet and the tally marks have been omitted from the following distributions.)

Letter	Frequency
A	31
E	7
H	6
I	7
K	9
L	8
M	8
N	11
O	13
P	1
U	14
W	2

13. The Hawaiian alphabet has only 12
    letters: A, E, I, O, U, and H, K, L, M,
    N, P, and W. The most frequently
    used letter is A.

14.

Letter	Frequency
A	13
E	3
H	0
I	3
K	0
L	0
M	0
N	0
O	4
P	0
U	5
W	0

15. That every Hawaiian word ends in a
    vowel.

16. Yes. All of the words end in vowels.

17.

Letter	Frequency
A	0
E	1
H	3
I	5
K	5
L	1
M	6
N	1
O	2
P	1
U	3
W	0

18. That Hawaiian words never begin
    with A or W.

19. No. There are words in the other
    paragraphs that begin with these
    letters.

Set III

1. These frequencies seem unusually
   large compared with the ones surrounding them.

2. The frequencies of people whose ages
   were reported as 25, 35, 45, 65, and 75
   follow a similar pattern. Except for
   these ages, frequencies of odd-
   numbered ages seem to be generally
   smaller than those of the adjacent
   even-numbered ages.

3. It would seem that many people in the
   census did not know their exact ages
   and gave only approximate ones,
   reporting their ages to the nearest 10
   years, 5 years, or 2 years.

Chapter 9, Lesson 2 (pp. 539-545)

Set I

1. HELP.

2. COME HERE AT ONCE.

3. E and T.

4. Q and Z.

5. 8.

6. $3 \times 8 = 24$.

7. 3.

8. $10 \times 3 = 30$.

9. No.

10. There are too few of each of these letters compared with their normal frequencies in English.

11. No.

12. There are too many of each of these letters compared with their normal frequencies in English.

13. Change the numbers of the letters on the sheet to the normal frequencies.

14. (The tally marks have been omitted from this distribution.)

Letter	Freq.	Letter	Freq.	Letter	Freq.
A	1	J	1	S	1
B	1	K	1	T	1
C	1	L	1	U	2
D	1	M	1	V	1
E	2	N	1	W	1
F	1	O	3	X	1
G	1	P	1	Y	1
H	1	Q	1	Z	1
I	3	R	1		

15. Because it contains every letter of the alphabet.

16. It appears less frequently.

17. It appears more frequently.

18. It would be difficult to decipher because the frequencies of the letters in it are not typical of ordinary English.

19. Q and Z.

20. Because they occur the least frequently in ordinary English.

Set II

1.
BFCON	ANYKF	I X KUS	I XHUC
I NTHE	SECON	DWORL	DWART
ONGFB	CNIAC	HCNAA	KSTN I
HEUNI	TEDS T	ATESS	OLVED
CONPH	BFYK I	NKMCO	NRHWH
THEMA	INCOD	E O FTH	E J A P A
FNANF	HTLCO	BAWSH	LN I HF
NES EN	AVYTH	ISPL A	YEDAN
BPWKU	CHFCU	KSNB F	CONHP
I MPOR	TANTR	OLE I N	THEAM
NUBYH	FTBYC	KULHC	CONIN
ER I CA	NVICT	ORYAT	THEDE
YBAB T	NDHCC	SNKMP	B I X HL
C I S IV	E B ATT	LEOFM	I DWAY

In the Second World War the United States solved the main code of the Japanese Navy. This played an important role in the American victory at the decisive battle of Midway.

2. To represent the space between each word.

3. No.

4. Because W's are used to represent spaces, spaces appear in place of W's when the message is decoded.

5. No.

Set III

The author of this story did not use any E's.

Chapter 9, Lesson 3 (pp. 548-553)

Set I

1. $\frac{45}{9} = 5$.

2. $\frac{48}{9} \approx 5.3$.

3. Thor.

4. No.

5. 5.

6. 2, 3, 3, 3, 4, 5, 5, 7, 16.

7. 4.

8. 7.

9. 3.

10. The median. (The mean determines which player wins; so it might be taken as the best measure of skill. However, it seems likely that Thor's 16 strokes on the last hole are a fluke and that he is in fact the best player. The median deemphasizes such flukes and is therefore the best measure.)

11. $162.81.

12. $\frac{\$162.81}{29} \approx \$5.61$.

13. $5.99.

14. $5.99.

15. The mean.

16. 89.

17. $\frac{192}{24} = 8$ centimeters.

18. No.

19. 0 centimeters (or, none).

20. 192 centimeters.

21. $\frac{32}{4} = 8$.

22. $\frac{432}{4} = 108$.

23. Each one is 100 more.

24. It is 100 more.

25. Subtract 100 from each number, average the numbers, and add 100 to the average.

26. 203.64.

Set II

1. $\frac{389}{25} = 15.56$.

2. 14.

3. No point spread occurs the most frequently.

4. The mode.

5. The mean.

6.

7. Line symmetry.

8. Each of them is 2.

9.

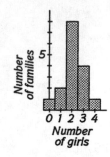

10. Mean, $\frac{34}{16} = 2\frac{1}{8}$; median 2; mode, 2.

11. It is not symmetrical.

12. It becomes 22. ($8 \times 21 = 168$;
    $168 + 30 = 198$; $\frac{198}{9} = 22$.)

13. 22.8. ($198 + 30 = 228$; $\frac{228}{10} = 22.8$.)

14. There were more people when Ole arrived.

15. His 102nd. ($11 \times 30 = 330$;
    $330 - 228 = 102$.)

Set III

1.

Class	Amount of money in billions
1	2
2	3
3	5
4	9
5	9

2.

Class	Amount of money in billions
1	3
2	3
3	5
4	7
5	10

3. The Grand Vizier's plan.

4. It is to the advantage of the poorest class to be averaged with the next-poorest class after the wealth of the latter has been increased.

5. The Grand Vizier's plan.

6. It is to the advantage of the richest class to be averaged with the next-richest class before the wealth of the latter has been reduced.

Chapter 9, Lesson 4 (pp. 557-563)

Set I

1. $128 - 114 = 14$ seconds.

2. $\frac{1,200}{10} = 120$ seconds.

3.

Time	Difference from mean	Square of difference
114	6	36
116	4	16
116	4	16
119	1	1
120	0	0
120	0	0
121	1	1
121	1	1
125	5	25
128	8	64

4. $\frac{160}{10} = 16$.

5. $\sqrt{16} = 4$.

6. $\frac{7}{10} = 70\%$.

7. $\frac{10}{10} = 100\%$.

8. $138 - 104 = 34$ seconds.

9. $\frac{1,200}{10} = 120$ seconds.

10.

Time	Difference from mean	Square of difference
104	16	256
116	4	16
116	4	16
119	1	1
120	0	0
120	0	0
121	1	1
121	1	1
125	5	25
128	18	324

11. $\frac{640}{10} = 64$.

12. $\sqrt{64} = 8$.

13. It increases.

14. It stays the same.

15. It increases.

16. $49 - 7 = 42$.

17. $\frac{1,150}{10} = 23$.

18. 33. (Scores between 12 and 34.)

19. $\frac{33}{50} = 66\%$.

20. 49. (Scores between 1 and 45.)

21. $\frac{49}{50} = 98\%$.

Set II

1. 33.5 centimeters.

2. $39 - 28 = 11$ centimeters.

3. 64.

4. 68%.

5. 68%.

6. 94%.

7. 96%.

8. 96%.

9. 100%.

10. 100%.

11. 100%.

12. It becomes more like a normal curve.

13. A normal curve.

14. 500.

15. $800 - 200 = 600$.

16. 100.

17. 34%.

18. 48%.

19. 50%.

20. 16%.

21. 16%.

22. 28,000 miles.

23. 17.

24. $\frac{17}{25} = 68\%$.

25. 24.

26. $\frac{24}{25} = 96\%$.

27. 25.

28. 100%.

29. The percentages are those expected for a normal distribution, but the histogram shows that the numbers do not form a normal distribution.

Set III

1. Curve C represents the newly minted quarters, curve B the coins after five years, and curve A the coins after ten years.

2. It decreases (from 7.5 grams to 7 grams to 6.5 grams.)

3. It increases.

Set I

1.

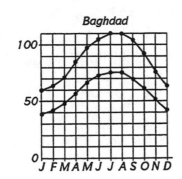

Baghdad

2.

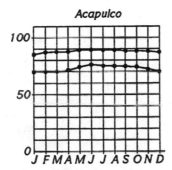

Acapulco

3.

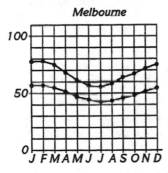

Melbourne

4. Acapulco.

5. Baghdad.

6. The graphs make it easier to see and compare the trends in temperature during the year (and they can be taken in at a glance).

7.

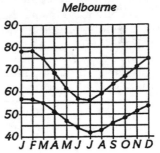

Melbourne

8. It makes the variation look more extreme.

9. The lower part of the graph is missing and the numbers representing the temperatures are spread farther apart.

10. No.

11. Almost four times as tall.

12. 96, 94, 93, 93, 94, 92, 98, 96, 91, 94.

13.

14. It is not as impressive: the tallest bar no longer looks so much taller than the others.

15. 1890.

16. No. The population of the United States in 1890 was smaller than in 1990. (The following table shows how the numbers compare.)

Year	Total population	Number of people less than 20 years old
1890	63,000,000	29,000,000
1990	249,000,000	72,000,000

17. It stops at 80% instead of 100%.

18. The bar looks taller than it should because of the part of the scale that is missing.

Set II

1. No.

2. The area of the larger bag is four times the area of the smaller one.

3. The volume of the larger bag is eight times the volume of the smaller one.

4. The bags used would be of equal size. Peter would have two of them and B.C. would have one.

5. The first set.

6. No. It makes the relative size of the USA audience appear to be bigger than it actually is.

7. It appears that the number of employees absent on November 10 was much larger than the number absent on the three preceding Mondays.

8. No.

9. It starts at 500 rather than 0.

10.

11. 1810, 1850, and 1950.

12. The decreases in population per square mile were due to increases in the land area of the United States.

13. The vertical scale starts at 21% rather than 0.

14. No.

15. It appears to have been about 21.8%.

16. (Student answer.)
Example graph:

Set III

1. Yes. Beginning at the age of 10, the dark-brown bar is taller than the light-brown bar. Beginning at the age of 16, the height of the dark-brown bar steadily increases with age.

2. Yes. The sum of the heights of each light-brown bar and corresponding dark-brown bar is 100.

3. So that a comparison of the percentages at each age is easier to make.

4. After childhood, people learn to associate large pupils with a happy appearance.

Chapter 9, Lesson 6 (pp. 576-581)

Set I

1. A group of items chosen to represent a larger group.

2. The people who vote in the election.

3. It could be too small. It might not be a random sample. The people surveyed might not tell the truth. They might change their minds between the time of the poll and the time of the election.

4. Not everyone living in the city has an equal chance of being included in the sample because not everyone has a telephone.

5. The people who have unlisted numbers.

6. Don't use a telephone book. Call random telephone numbers.

7. $64,651 - $64,568 = $83.

8. $5,000 - $1,000 = $4,000.

9. Taking a sample of the freight bills would benefit the company more. The money saved by taking the sample would be greater than the money lost by trusting it.

10. There are too many songs broadcast (more than a billion each year!)

11. They take a random sample of broadcasts and base the payments on how frequently each composer's songs are played on them.

12. Perhaps 80% of all pedestrians wear dark clothes and 20% wear light-colored clothing.

13. A survey of the clothing worn by all pedestrians.

Set II

1. Not necessarily.

2. It was too small. (Had there been many cases among the children not inoculated and none among those inoculated, the vaccine could be thought to be effective. However, if both groups had been free of the disease, no conclusions about the effectiveness of the vaccine could be drawn.)

3. $\frac{10}{50} = \frac{1}{5}$.

4. $\frac{1}{5}$.

5. 500.

6. Perhaps the fish caught in the first sample were less intelligent than the rest of the fish, so that their chance of being caught again is greater. Perhaps the fish caught in the first sample became wary of being caught again, so that the chance of it happening again is less.

7. The drivers polled might be reluctant to admit to the poll-takers (and their passengers) that their cars are not completely paid for.

8. It would make the figure reported too large.

9. That most Coca-Cola drinkers prefer Pepsi.

10. That people prefer the letter M to the letter Q.

11. Most people preferred the Coca-Cola served in the glasses marked M.

12. (One possible answer.) Serve the colas in unmarked glasses.

13. $1 \times 17 + 2 \times 23 + 3 \times 22 + 4 \times 15 + 5 \times 9 + 6 \times 8 + 8 \times 6 = 330$; $\frac{330}{100} = 3.3$.

14. Because it is likely that more than one child could be attending the camp from a given family, many families could have been reported more than once. For example, it is possible that all six of the children who reported eight children in their families are from the same family.

Set III

1. The empty envelopes labeled "research study."

2. The envelopes seeming to contain money.

3. To get an idea of how many envelopes might not be returned for reasons other than dishonesty. For example, some might be picked up by children or be blown out of view by the wind.

4. They were counted as if they had not been returned.

5. The number of envelopes seeming to contain money that were returned varies with the area of the city; the fewest were returned from the poor neighborhoods and the most were returned from the wealthy sections.

6. An envelope seeming to contain money would be a greater temptation to a poor person than to a rich one. (One of the psychologists who did the experiment said about the results, "The only theory of dishonest behavior that may have been supported by our study is that people act according to their needs.")

Chapter 9, Review (pp. 584-590)

Set I

1. January 1855.

2. October 1854-January 1855.

3. Four.

4. The median.

5. The mean (or, mode).

6. Yes.

7. Numbers along the axes.

8. No.

9. It seems strange that the graph consists of line segments that suddenly change direction. A more accurate graph would probably consist of curves rather than straight line segments.

10. To save time and money.

11. A normal curve.

12. 200 centimeters.

13. 5 centimeters.

14. 68%.

15. 96%.

16. 100%.

17. 500,000.

18. None.

19. Oregon.

20. No. California is larger than Oregon. (46 million acres in California and 33 million acres in Oregon are federally owned.)

Set II

1. The second graph (for 800 meters). The first interval of time on the y-axis is 50 seconds and the rest of the intervals are 10 seconds each, which makes it seem as though the runner took almost twice as long as the skater.

2. They have different time scales.

3. 153.

4. 15.3.

5. The mode. More people got it than any other measurement.

6. $\frac{244.3}{7}$ = 34.9 centimeters.

7. The median and the mode.

8.

Time in minutes	Tally marks	Frequency
71-80	I	1
81-90		0
91-100	ⅢⅠ I	6
101-110	ⅢⅠ III	8
111-120	ⅢⅠ ⅢⅠ I	11
121-130	III	3
131-140	I	1

9.

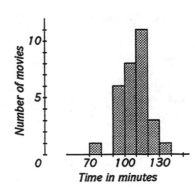

Time in minutes

10. $\frac{21}{30}$ = 70%. (21 of the movies are between 99 and 123 minutes long.)

11. $\frac{28}{30}$ ≈ 93%. (28 of the movies are between 87 and 135 minutes long.)

12. Yes. For a normal curve, the expected percentages are 68% and 96%.

13. Some of the people in the sample who said they voted did not.

14. The sample might be too small. It might not be random.

15. The numbers of readers seem to have been used in determining the heights of the trucks.

16. Yes. Many people looking at the graph would probably compare the areas or volumes of the trucks instead.

Set III

1. E.

2. I and T.

3. A, N, and S.

4. The frequencies are among the highest of the letters of the alphabet.

5. J, Q, and Y.

6. They are very low.

7. C, O, P, X, and Z.

8. O.

9. Because its frequency is high.

Chapter 9, Further Exploration

Lesson 1 (pp. 591-593)

1. a) (Example.)

Greatest number of cards in any suit	Frequency
3	7
2	53
1	40

b) $\frac{52}{52} \times \frac{12}{51} \times \frac{11}{50} \approx 5\%.$

c) $\frac{52}{52} \times \frac{39}{51} \times \frac{26}{50} \approx 40\%.$

d) 55%.

e) (Student answer.)

2. a)

Number of times per 1,000 words	48 essays by Hamilton	50 essays by Madison	12 disputed papers
1.0-2.9	4%	0%	0%
3.0-4.9	15%	0%	0%
5.0-6.9	25%	10%	17%
7.0-8.9	38%	14%	8%
9.0-10.9	8%	16%	17%
11.0-12.9	10%	32%	33%
13.0-14.9	0%	12%	17%
15.0-16.9	0%	10%	8%
17.0-18.9	0%	6%	0%

b)

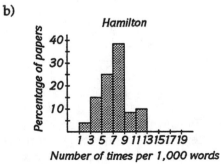

Hamilton

Percentage of papers

Number of times per 1,000 words

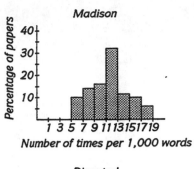

Madison

Percentage of papers

Number of times per 1,000 words

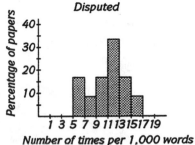

Disputed

Percentage of papers

Number of times per 1,000 words

c) Madison.

d) They seem to indicate that the disputed papers were written by Madison. The histogram for the disputed papers looks much more like the one for Madison than like the one for Hamilton.

Lesson 2 (pp. 593-595)

1. a) 32.   f) 71.
   b) 52.   g) 20.
   c) 16.   h) 9.
   d) 57.   i) 46.
   e) 43.   j) 54.

   k) The fingers remain on the home row much more of the time.

   l) The distribution of work between the two hands is somewhat more even, with the right hand rather than the left hand getting more of the work. This is an improvement because most people are right-handed.

2. a) IF YOUR PARENTS DIDN'T HAVE CHILDREN, THERE'S A GOOD CHANCE THAT YOU WON'T HAVE ANY.
—CLARENCE DAY

b) I LIKE LONG WALKS, ESPECIALLY WHEN THEY ARE TAKEN BY PEOPLE WHO ANNOY ME.
—FRED ALLEN

c) THE BEST WAY TO KEEP CHILDREN HOME IS TO MAKE THE HOME ATMOSPHERE PLEASANT—AND LET THE AIR OUT OF THE TIRES.
—DOROTHY PARKER

Lesson 3 (pp. 595-596)

1. a) It is 1 for German and Japanese and 2 for Arabic.

b) German, 1; Japanese, 2; Arabic, 2.

c) $\dfrac{71 \times 1 + 19 \times 2 + 7 \times 3 + 2 \times 4 + 1 \times 5}{100} =$ 1.43 syllables per word.

d) German:
$\dfrac{56 \times 1 + 31 \times 2 + 9 \times 3 + 3 \times 4 + 1 \times 5}{100} =$ 1.62 syllables per word.

Japanese:
$\dfrac{36 \times 1 + 34 \times 2 + 18 \times 3 + 9 \times 4 + 2 \times 5 + 1 \times 6}{100} =$ 2.10 syllables per word.

Arabic:
$\dfrac{23 \times 1 + 50 \times 2 + 22 \times 3 + 5 \times 4}{100} =$ 2.09 syllables per word.

2. a) (Student record.)

b) (Example.)

Number of rolls	Frequency	Number of rolls	Frequency
6	1	18	0
7	0	19	0
8	3	20	1
9	1	21	2
10	0	22	0
11	2	23	1
12	3	24	0
13	2	25	0
14	2	26	0
15	1	27	0
16	3	28	1
17	2		

c) (Example.) The median is 14.

d) (Example.) The mean is approximately 14.4.

The theoretical number of rolls necessary is 14.7. The probability that we get one of the six numbers on the first roll is 1. The probability of getting a number different from the first one is $\dfrac{5}{6}$, and so it will take an average of $\dfrac{6}{5}$ rolls to get it. The probability of getting a number different from the first two is $\dfrac{4}{6}$, so it will take an average of $\dfrac{6}{4}$ rolls to get it. To get all six numbers, then, takes an average of $1 + \dfrac{6}{5} + \dfrac{6}{4} + \dfrac{6}{3} + \dfrac{6}{2} + \dfrac{6}{1} =$ $1 + 1.2 + 1.5 + 2 + 3 + 6 = 14.7$ rolls.

Lesson 4 (pp. 596-597)

1. a) 250 days and 282 days. (If the lengths of pregnancy are normally distributed, 68% of them would be expected to be within one standard deviation or less of the mean.)

b) 234 days and 298 days. (96% would be expected to be within two standard deviations of the mean.)

c) 218 days and 314 days. (Almost 100% would be expected to be within three standard deviations of the mean.)

d) 0.3%. $(310 - 266 = 44; \frac{44}{16} = 2.75.$

According to the table on page 596, 99.4% of the numbers in a normal distribution are within 2.75 standard deviations of the mean. So the remaining 0.6% are more than 2.75 standard deviations from it; 0.3% above and 0.3% below.)

e) About three. (0.3% of 1,000 = 3.)

f) Yes. A pregnancy of 310 days is very unusual, but certainly possible.

2. a) 246 grams.

b)

Weight in grams	Frequency
230-234	1
235-239	4
240-244	8
245-249	8
250-254	3
255-259	1

c)

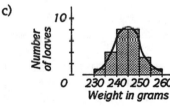

d) It looks like a normal curve.

e) About 245 grams.

f)

Weight in grams	Frequency
230-234	0
235-239	0
240-244	0
245-249	0
250-254	12
255-259	3

g)

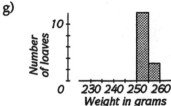

h) Instead of increasing the average weight of the loaves to 250 grams, the baker was picking out loaves that weighed at least 250 grams to sell to the student.

i) The histogram is no longer symmetrical. It is reasonable to think that the variation in the weights of the loaves should still suggest a normal curve.

Lesson 5 (pp. 598-600)

1. a) The line that starts out higher.

b) They increased in the first three weeks, decreased in the three weeks following the price change, and then began increasing again.

c) They steadily increased.

d) Perhaps an increase in the initial price of a product causes a buyer to consider it overpriced and to stop buying it. Perhaps the higher the price paid for a product at the beginning, the more a buyer will value it and continue to purchase it.

317

e) (The data do not tell us. This is a matter of judgment. Different students will have different answers.)

2. a)

Grade level	Percentage of male applicants hired	Percentage of female applicants hired
K-6	5	10
7-9	14	30
10-12	40	90
Total	30	18

b)

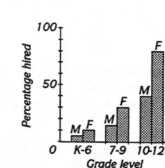

c) It seems to show that the district discriminates in favor of women. The percentages of female applicants hired at each level are greater than the corresponding percentages of male applicants hired. (This is an example of Simpson's paradox, and it shows that the conclusions drawn from the same data depend on how the data are broken down into subgroups. There is no absolute answer to whether these data reveal bias in hiring. Furthermore, nonstatistical investigations would be needed to settle the question of bias. See the Miss Lonely Hearts paradox in Martin Gardner's *Aha! Gotcha* for further discussion.)

d)

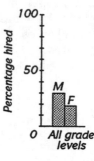

e) It seems to show that the district discriminates in favor of men. The percentage of male applicants hired at all levels combined is greater than the percentage of female applicants hired.

Lesson 6 (p. 600)

1. (Three random samples, each consisting of ten pages of the book, indicate that the book contains approximately 138,000 words.)

2. The missing stockings were not being stolen. They were never made. The annual output of the company was estimated from a sample that was too small (it consisted of just one person and the amount of material used in making the stockings varied from person to person) and not random (the person was the company's best worker.) She used less material per stocking than the average worker, and so the average worker made fewer stockings from a given amount of material than she did.

Chapter 10, Lesson 1 (pp. 604-609)

Set I

1. A and C.

2. Topologically equivalent.

3. No.

4. The first figure consists of a single curve; the second figure consists of two separate curves.

5. C, G, I, J, L, M, N, S, U, V, W, and Z.

6. D and O.

7. K and X.

8. P and Q.

9. A and R.

10. E, F, T, and Y.

11. Yes.

12. Yes.

13. Yes.

14. No. Region 4 does not share some of its border with region 5.

15. No. (Such a map would contradict the four-color theorem.)

Set II

1. The figure is a simple closed curve because it is possible to start at any point and travel over every other point exactly once before returning to the starting point.

2. The cardioid and the octagon.

3. The cardioid and the octagon.

4. Outside.

5. Line 1 crosses the curve 2 times, line 2 crosses it 6 times, and line 3 crosses it 4 times.

6. Inside.

7. Line 4 crosses the curve 3 times, line 5 crosses it 5 times, and line 6 crosses it 7 times.

8. If the line crosses the curve an *odd* number of times, the point is *inside* the curve. If it crosses an *even* number of times, the point is *outside*.

9. Point C is inside, point D is outside, and point E is inside.

10. (Drawing by student.)

11. A simple closed curve.

12. Electricity to house B.

13. Electricity is inside the curve and house B is outside it.

14. The puzzle has no solution.

15. One possible answer:

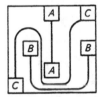

16. This figure does not have such a circuit because any line connecting A to A separates B from B.

17. One possible answer:

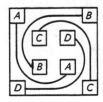

319

Set III

1. They are topologically equivalent.

2. No. The parts of the picture farthest from the base of the cone are twisted and stretched the greatest amount.

3. It turns it inside out.

Chapter 10, Lesson 2 (pp. 611-616)

Set I

1. Yes.

2. The degree of a vertex is the number of edges that meet at it.

3. 2.

4. Yes.

5. 2.

6. 2.

7.

Network	Number of even vertices	Number of odd vertices	Can the network be traveled?
A	4	0	Yes
B	2	2	Yes
C	0	4	No
D	2	2	Yes
E	0	4	No
F	4	0	Yes
G	0	4	No
H	2	2	Yes
I	4	0	Yes

8. All of them are odd.

9. A, B, D, and E.

10. None of them.

11. C.

12. C.

13–15. (Student drawings.)

16. It has an extra vertex (at Lanai).

17. It has an extra edge (connecting Kauai and Maui.)

18. A and B.

19. No.

20. Yes.

Set II

1. The degrees of the vertices.

2. 1.

3. 4.

4–13. (Student drawings.)

14.

Network	Number of even vertices	Number of odd vertices	Can the network be traveled?
4	9	0	Yes
5	0	6	No
6	3	2	Yes
7	6	2	Yes
8	4	4	No
9	10	0	Yes
10	1	6	No
11	1	2	Yes
12	5	4	No
13	9	0	Yes

15. Yes.

16. No.

17. Yes.

18. No.

19. A network can be traveled if it has no more than two odd vertices.

20. (Student drawing.)

21. At points C and E (or vice versa).

22. Their degree is odd.

Set III

1.

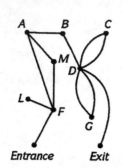

2. Yes. The path would go through rooms F, M, A, B, and D.

3. No.

Chapter 10, Lesson 3 (pp. 618-624)

Set I

1. A continuous path that passes along every edge of a network exactly once.

2. A.

3. A.

4. C.

5. C.

6. B.

7. B.

8. A network has an Euler path that returns to its starting point if it has no odd vertices.

9. A network has an Euler path that does not return to its starting point if it has two odd vertices.

10. A network does not have an Euler path if it has more than two odd vertices.

11. Yes.

12. It has no odd vertices.

13. (Student drawing.)

14. No.

15. Yes; you would end at A.

16. Yes; you would end at B.

17. It has no odd vertices.

18. Network B, 3 trips; network C, 4 trips; network D, 5 trips.

19.

Network	Number of odd vertices	Least number of trips
A	4	2
B	6	3
C	8	4
D	10	5

20. The number of trips necessary to travel a network is half the number of odd vertices that it has.

21.

Network	Number of regions	Number of vertices	Number of arcs
A	3	2	3
B	1	6	5
C	2	3	3
D	1	8	7
E	3	1	2
F	5	5	8
G	6	8	12
H	9	5	12
I	4	13	15

22. $R + V = E + 2$. (Or, $R + V - 2 = E$.)

23. 5 regions, 18 vertices, and 21 edges.

24. Yes; $5 + 18 = 21 + 2$.

Set II

1.

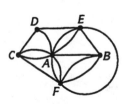

2. Vertices D and E are odd; the other vertices are even.

3. Yes. No. (It must begin and end either on D or on E.)

4. A and D.

5. A and B.

6. Yes.

7. Yes.

8. A Hamilton path.

9. Because the person responsible for the newspaper racks wants to travel to every vertex of the network, not along every edge.

10. Yes, the second map does. (Example.)

11. No. It has more than two odd vertices.

12. (Example.)

3. 335. (The lengths of the three possible paths are:

$$110 + 100 + 80 + 60 = 350,$$
$$110 + 55 + 80 + 120 = 365,$$
$$60 + 55 + 100 + 120 = 335.)$$

Set III

1. (Student drawing.)

2. Three.

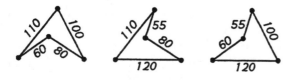

Chapter 10, Lesson 4 (pp. 627-633)

Set I

1. 15.

2. 14.

3. 17.

4. 16.

5. The number of edges is one less than the number of vertices.

6. 1, 2, 3, 4, 5, and 7.

7.

Digit	Number of vertices	Number of edges	Is it a tree?
1	2	1	Yes
2	6	5	Yes
3	6	5	Yes
4	5	4	Yes
5	6	5	Yes
6	6	6	No
7	3	2	Yes
8	6	7	No
9	6	6	No
0	4	4	No

8. Yes.

9. $V = E + 1$. (Or, $V - E = 1$.)

10. (Student drawing. Such a tree is not possible.)

11. Yes; $20 = 19 + 1$.

12. No; $17 \neq 18 + 1$.

13. 2.

14. Yes.

15. Yes.

16. 6.

17. 19.

18. 18.

19. 4.

20. 7.

Set II

1. 1.

2. 2.

3. 3.

4. 1 and 3.

5. 2 and 4.

6. 5 and 9.

7. 6 and 7.

8. 8 and 10.

9. K.

10. H.

11. M.

12. J.

13. I.

14. L.

15. No. (Trees B and D, for example, both have a diameter of 2 but are not topologically equivalent.)

16. Yes.

17.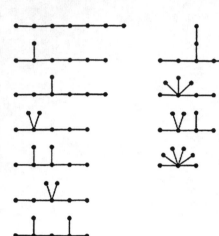

5. Yes.

6.

Hydrocarbon	Number of carbon atoms	Number of hydrogen atoms
Methane	1	4
Ethane	2	6
Propane	3	8
Butane	4	10
Pentane	5	12

7. 8.

8. 18.

Set III

1. 4.

2. 1.

3. *Propane*

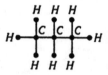

Butane

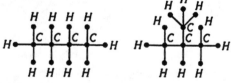

4. *Pentane*

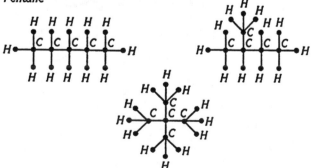

Chapter 10, Lesson 5 (pp. 636-640)

Set I

1. One.

2. Two.

3. Two separate loops.

4. Two separate loops.

5. One is twice as wide as the other.

6. They have equal lengths.

7. Twice.

8. Both.

9. Just one.

10. You would paint the entire strip.

11. The result is just one band, rather than two as might be expected.

12. One.

13. Two.

14. No. A Moebius strip has only one side.

15. Two interlocking bands.

16. Two interlocking bands.

17. They have equal widths.

18. One is twice as long as the other.

19. The shorter band.

20. Cutting a Moebius strip along its center produces just one band, rather than two.

Set II

1. Yes.

2. Two.

3. Two interlocking bands.

4. They have equal widths and equal lengths.

5. Two interlocking bands.

6. One is twice as wide as the other. They have equal lengths.

7. One.

8. A single band with a knot in it.

9. Two interlocking bands. The longer band has a knot in it and the shorter band passes through the knot.

10. Two.

11. Two interlocking bands.

12. A single band with a knot in it.

13. A band with an odd number of half-twists in it has one side; a band with an even number of half-twists in it has two sides.

14. Two interlocking bands.

15. A single band with a knot in it.

Set III

1. No.

2. The strip becomes untangled.

Chapter 10, Review (pp. 642-646)

Set I

1.

2.
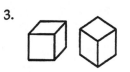

3.

4. A vertex.

5. An edge.

6. The hall.

7. The kitchen.

8. Yes, because it does not contain any closed curves.

9. 4.

10. A Moebius strip.

11. One.

12. A single band, twice as long.

13. Yes.

14. No.

15. The second map has more than two odd vertices.

16. No. The route must begin at either C or E.

Set II

1. 13.

2. 15.

3. 4.

4. Yes. $(4 + 13 = 15 + 2.)$

5–9. (Drawings by student.)

10.

Network	Number of even regions	Number of odd regions	Does the network have a path?
5	2	0	Yes
6	5	1	Yes
7	2	3	No
8	1	2	Yes
9	0	4	No

11. Whether or not the network has more than two odd regions. If it does not, there is such a line.

12. No. It has three odd regions.

Set III

1. 2-3, 2-5, 3-6, 4-5, and 4-6.

2. No.

3. Six.

Chapter 10, Further Exploration

Lesson 1 (pp. 647-649)

1. a) Yes. Three rings linked in this way are shown in this figure.

   b) They also are linked together without any two of them being linked with each other.
   c) They were carved out of a single piece of wood.

2. a) The numbers assigned to the W's are 16, 8, 4, 2, and 1. Which of these numbers is assigned depends on the position of the W in the rows. The other letters are assigned the number 0.
   b) Each number in the file classification is determined by adding 1 to the sum of the numbers in the corresponding row.
   c) None of the fingerprints is a whorl. (If the file classification is $\frac{1}{1}$, the numbers must be $\frac{00000}{00000}$)
   d) The prints of the middle finger and index finger of the person's right hand are whorls. (If the file classification is $\frac{17}{9}$, the numbers must be $\frac{160000}{08000}$.)

Lesson 2 (pp. 649-650)

1. a)

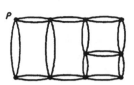

   b) Yes. One such path is shown in the figure below.

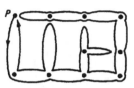

   c) All of them are even.
   d) Yes. If the carrier delivers mail on both sides of each street, each street is represented by two edges so that each vertex of the network that represents the territory must be even. A network all of whose vertices are even can always be traveled in a path that ends where it begins.

2. a) No. It can last no longer than 5 moves.
   b) It can last no longer than 8 moves.
   c) It can last no longer than 11 moves.
   d) It can last no longer than 29 moves. (In general, a game that starts with n points must end in $3n - 1$ moves or less.) (For many other fascinating games, several of them topological, see *Winning Ways*, volumes 1 and 2, by Berlekamp, Conway, and Guy, Academic Press, London, 1982.)

Lesson 3 (pp. 650-651)

1. a) (Student drawing.)
   b) The network can be drawn only if you begin at one of two of its 76 vertices.
   c) To draw the network, begin at either of the two odd vertices; one is the middle vertex at the top and the other is the middle vertex at the far right.

2. a) No. The network has four odd vertices, and so it cannot be traveled in a single trip.
   b) The path shown in this figure contains only two retraced sections, both of which are very short.

Lesson 4 (pp. 651-652)

1. a) The tree diagrams below show the two ways in which the four people may be related.

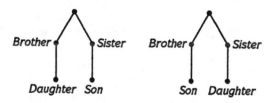

2. a) (A communications network with a total risk of 23 is shown below.)

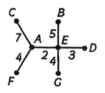

b) Yes. If it is not a tree, one or more of its edges can be removed to reduce the total risk and yet leave a communication path between every pair of agents.
c) (Student answer.)
d) (Student answer. A logical method to begin the network is to connect the agent pairs with the smallest risk factors first.) (This is a combinatorial optimization problem. Operations research methods are used to attack such problems systematically.)

Lesson 5 (pp. 652-654)

1. a) A chain of three interlocking loops.
   b) One long band.

2. a) A large square.
   b) A large square identical with the one produced in Part 1.
   c) Two separate pieces shown in the figure below.

d) Two interlocking pieces, of which one is a large square and the other is a small Moebius strip.

SOURCES OF ILLUSTRATIONS